D1356025

Transnational Politics of the Environment

Global Environmental Accord: Strategies for Sustainability and Institutional Innovation
Nazli Choucri, editor

Nazli Choucri, editor, *Global Accord*

Peter M. Haas, Robert O. Keohane, and Marc A. Levy, editors, *Institutions for the Earth*

Ronald B. Mitchell, *Intentional Oil Pollution at Sea*

Robert O. Keohane and Marc A. Levy, editors, *Institutions for Environmental Aid*

Oran R. Young, editor, *Global Governance*

Jonathan A. Fox and L. David Brown, editors, *The Struggle for Accountability*

David G. Victor, Kal Raustiala, and Eugene B. Skolnikoff, editors, *The Implementation and Effectiveness of International Environmental Commitments*

Mostafa K. Tolba, with Iwona Rummel-Bulska, *Global Environmental Diplomacy*

Karen T. Litfin, editor, *The Greening of Sovereignty in World Politics*

Edith Brown Weiss and Harold K. Jacobson, editors, *Engaging Countries*

Oran R. Young, editor, *The Effectiveness of International Environmental Regimes*

Ronie Garcia-Johnson, *Exporting Environmentalism*

Lasse Ringius, *Radioactive Waste Disposal at Sea*

Robert G. Darst, *Smokestack Diplomacy*

Urs Luterbacher and Detlef F. Sprinz, editors, *International Relations and Global Climate Change*

Edward L. Miles, Arild Underdal, Steinar Andresen, Jørgen Wettestad, Jon Birger Skjærseth, and Elaine M. Carlin, *Environmental Regime Effectiveness*

Erika Weinthal, *State Making and Environmental Cooperation*

Corey L. Lofdahl, *Environmental Impacts of Globalization and Trade*

Oran R. Young, *The Institutional Dimensions of Environmental Change*

Tamar L. Gutner, *Banking on the Environment*

Liliana B. Andonova, *Transnational Politics of the Environment*

Transnational Politics of the Environment

The European Union and Environmental Policy in Central and Eastern Europe

Liliana B. Andonova

The MIT Press
Cambridge, Massachusetts
London, England

Set in Sabon by SNP Best-set Typesetter Ltd., Hong Kong. Printed and bound in the United States of America. Printed on Recycled Paper.

Library of Congress Cataloging-in-Publication Data

Andonova, Liliana B.
Transnational politics of the environment : EU integration and environmental policy in Central and Eastern Europe / Liliana B. Andonova.
p. cm.—(Global environmental accord: strategies for sustainability and institutional innovation)
Includes bibliographical references and index.
ISBN 0-262-01206-5 (hc.)—ISBN 0-262-51179-7 (pbk.)
1. Environmental policy—Europe, Central—International cooperation.
2. Environmental policy—Europe, Eastern—International cooperation.
3. Environmental policy—European Union countries. I. Title. II. Global environmental accords

GE190.C36A53 2003
363.7'056'0943—dc21 2003051210

to my family

Contents

Acknowledgments ix

Introduction 1

I Chemical Safety Regulations: The Harmonizing Influence of International Markets and Institutions 35

1 Regulatory Pressure and the Chemical Industry 39

2 Chemical Safety Policies in the Czech Republic, Poland, and Bulgaria 65

II European Regulations to Combat Air Pollution 85

3 The Czech Republic: Early Adaptation 95

4 Poland: The Bargain of the Electricity Industry 123

5 Bulgaria: Harmonization without Implementation 153

Conclusion 183

Appendix: List of Interviews 197

Notes 203

Bibliography 225

Index 253

Acknowledgements

When I was little, I often wondered how books get written and how airplanes fly. Now, I believe I have some answers to the first question. The writing of this book involved a lot of personal inspiration, but also tremendous support from colleagues, friends, and family. I would like to express my gratitude to my academic advisers at Harvard University, Robert Bates, Robert Keohane, and Lisa Martin, for their advice and commitment, for challenging me, and for inspiring creative thinking and high standards. I am also indebted to Marc Busch, Bill Clark, Grzegorz Ekiert, and Jeff Frieden for many insightful discussions. I thank Lewis Gilbert and Peter Bearman for supporting my book completion endeavor as a postdoctoral research scholar at the Earth Institute and the Institute for Social and Economic Research and Policy at Columbia University. My gratitude goes to many colleagues and friends—David Cash, Matt Gershoff, Tom Garvey, Heather Greene, Larry Hamlet, Judit Kelley, Mariana Panuncio, Claudia Scholz, Ken Scheve, Oxana Shevel, Josh Tucker, Mike Tomz, and Inger Weinbust—who read drafts and provided comments on earlier versions of the manuscript. I am indebted to Robert Keohane for encouraging me to publish the book in a timely manner, and to Clay Morgan for guiding the publication process at The MIT Press. I also thank Ronald Mitchell and Tamar Guter for providing thoughtful and valuable comments that guided some of the final revisions. I thank John Attkinson and Philip Gunev for research assistance and Charlotte Forbes for editing assistance. With this book, I also honor the memory of Ronie Garcia-Johnson, a dear friend and colleague, who shared her outstanding work, enthusiasm, and advice with me, and helped tremendously the timely publication of this research.

The research for this project was supported by the Program for the Study of Germany and Europe of the Minda de Gunsburg Center for European Studies, by the World Policy Institute, by the Joseph Fisher Price of Resources for the Future, by the Mellon Dissertation Completion Fellowship, and by summer travel grants from the Harvard Davis Center for Russian and East European Studies, the Harvard Weatherhead Center for International Affairs, the Columbia Earth Institute, and the Columbia Institute for Social and Economic Research and Policy. I am much obliged for this invaluable support. I would also like to recognize the help of everyone with whom I worked in Bulgaria, the Czech Republic, Poland, and Brussels during my extensive travels, without whose assistance this book would have been impossible.

This book is dedicated to my family, which has been a source of strength and personal support for my academic pursuits. I owe special gratitude to my husband, Tihomir Andonov, for reading more drafts than can be counted and for supporting me every step of the way. I thank my parents, Saba and Nikolay Bochev, for the encouragement to always strive for higher standards. I thank my father-in-law, Andon Andonov, for taking care of the family while I was finishing this book. I thank my daughter Daniela for her lovely, inspiring smile, and for checking regularly if mama is still writing, and to my newborn daughter Nicolena for patiently waiting until all the finishing touches of the book were made.

Transnational Politics of the Environment

Introduction

The objective of European Union[1] accession dominates the environmental policy agenda of Central and East European countries. Shortly after the transition to democracy, Bulgaria, the Czech Republic, Estonia, Hungary, Latvia, Lithuania, Poland, Romania, Slovakia, and Slovenia declared their objective to rejoin "the family of Europe."[2] One major condition for achieving the strategic goal of EU membership, however, is the adoption of the Acquis Communautaire ("the Acquis," for short)—the body of EU regulations, including costly environmental standards. This book examines the influence of EU integration on environmental politics in Central and Eastern Europe. It asks how are the costs and benefits of integration and environmental regulation distributed among producers, consumers, and taxpayers in different states? What kind of coalitions form as a result? What is the role of domestic institutions in shaping responses to EU pressures? And what are the implications for the scope, implementation, and effectiveness of environmental policies?

In order to illuminate these important questions, the book examines the effect of the dual force of EU markets and institutions on domestic political processes. It challenges the existing focus in the EU enlargement literature on inter-governmental cooperation, and it highlights the roles of industries, international norms, and domestic institutions in linking international and domestic politics. The explanatory power of the framework is evaluated in detailed case studies of the development of chemical safety and air pollution policies in Bulgaria, the Czech Republic, and Poland during the first decade of post-communist reforms (1990–2000).

The study contributes to several fields in international relations and comparative politics. To students of post-communist transitions and European integration, it offers a method for comparative analysis of the effects of regional integration on domestic politics and policy choice. To the environmental policy literature, the book adds a framework for understanding the internationalization of domestic environmental politics. The book also contributes to political economy studies more broadly by extending the open-economy model to environmental regulations, shifting the focus away from industrialized states to emerging markets, and integrating insights from interest-based and institutional theories.

This introduction begins the journey into the transnational history of post-communist environmental politics in Central and Eastern Europe. It provides a brief background of EU enlargement and its environmental conditions, and it elaborates a theoretical approach to understating the interaction between regional and domestic environmental politics. After explaining the rationale for selecting the cases of chemical and air pollution regulations, it provides a road map to the structure of the book.

The Challenges of EU Integration and Environmental Harmonization

Membership in the EU is of paramount strategic and symbolic significance for Central and East European countries. It symbolizes the completion of the coveted "return to Europe" from the Cold War dominance of Russia. EU accession guarantees greater external security, economic prosperity, and consolidation of the newly established democracies. It is not surprising, therefore, that the process of regional integration has permeated almost all aspects of the social and political life in Central and Eastern Europe since the early 1990s.[3] This process involves two parallel trends: regional market integration and political integration toward EU accession.

The market integration between Eastern and Western Europe proceeded rapidly after the democratic changes of 1989. In the early 1990s, Central and East European states signed association agreements with the EU that liberalized trade almost completely, with the exception of agriculture and the maintenance of safeguard and anti-dumping measures

Table I.1
Central and East European countries' exports and imports to EU as percentages of their total exports and imports, 2001. Source: Commission of the European Communities 2002.

	Exports	Imports
Bulgaria	54.8	49.4
Czech Republic	68.9	61.8
Estonia	69.4	56.5
Hungary	74.3	57.8
Latvia	61.2	52.6
Lithuania	47.8	44
Poland	69.2	61.4
Romania	67.8	57.3
Slovakia	59.9	49.8
Slovenia	62.2	67.7

that could be evoked to protect sensitive sectors such as steel, coal, chemicals, and textiles.[4] The EU became the most important trading partner of these countries as its share in their exports and imports expanded to more than 50 percent by 2001 (table I.1). Foreign direct investment (FDI) also increased rapidly, especially in Hungary, Poland, and the Czech Republic—the countries most advanced in reforming their economies (table I.2, figure I.1). EU member states are the leading source of FDI in Central and Eastern Europe. In 2001, for example, the EU accounted for 77 percent of FDI in Bulgaria, 88 percent of FDI in the Czech Republic, and 64 percent of FDI in Poland.[5]

The parallel process of political integration toward EU membership proceeded considerably more slowly than regional market integration. At its 1993 Copenhagen Summit the EU formally committed to granting membership to Central and East European candidates and established three broad criteria for accession: stability of democratic institutions, functioning market economy, and ability to apply EU legislation.[6] The literature that describes the political integration of the two parts of Europe highlights the multiple political and institutional obstacles to the eastward enlargement of the Union.[7] Internally, there is differentiation of interests between member states that anticipate economic or political benefits from enlargement (Germany, Denmark, Sweden, Finland, the United Kingdom) and those (Portugal, Spain, Greece,

Table I.2
FDI trends in Central and Eastern Europe, 1990–2000 (million current US$). Source: World Bank 2002.

	1990	1991	1992	1993	1994	1995	1996	1997	1998	1999	2000
Bulgaria	4	56	42	40	105	90	109	505	537	806	1,002
Czech Republic	207	600	1,103	654	878	2,568	1,435	1,286	3,700	6,313	4,583
Estonia			82	162	214	202	150	266	581	305	387
Hungary		1,462	1,479	2,350	1,144	4,519	2,274	2,167	2,037	1,977	1,692
Latvia			29	45	215	180	382	521	357	348	407
Lithuania				30	31	73	152	355	926	487	379
Poland	89	291	678	1,715	1,875	3,659	4,498	4,908	6,365	7,270	9,342
Romania		40	77	94	341	419	263	1,215	2,031	1,041	1,025
Slovakia				199	270	236	351	174	562	354	2,053
Slovenia			111	113	128	177	194	375	248	181	176

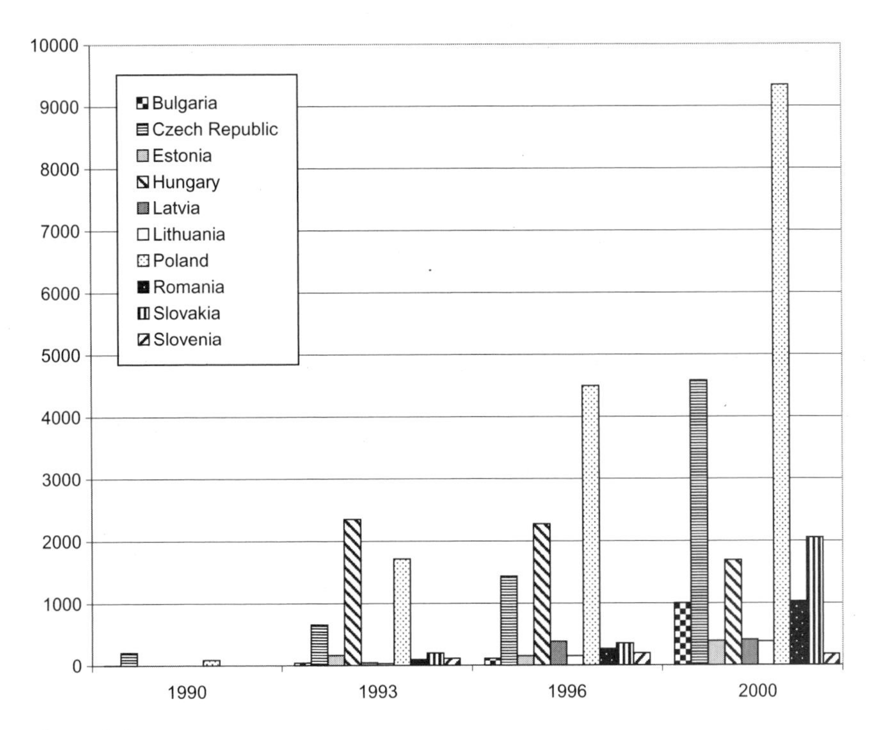

Figure I.1
FDI trends in Central and Eastern Europe, 1990–1999 (million current US$). Source: World Bank 2002.

Ireland) that fear that enlargement will redistribute EU resources toward the East, to their overall disadvantage. Opposition to enlargement also comes from powerful societal interests in the Union, such as farmers concerned with the future of the Common Agricultural Policy, sensitive industries, workers apprehensive of a possible inflow of cheap labor, and taxpayers in countries (e.g., Germany and the Netherlands) that are major contributors to the EU budget. The accession of ten new members with considerably lower per capita income than the EU average (figure I.2), furthermore, necessitates reform of EU institutions and of such redistributive mechanisms as the Structural Funds and the Common Agricultural Policy.

Despite these internal hurdles, however, the EU started formal accession negotiations with the ten Central and East European candidates in 1998–99. The Treaty of Nice (2001) introduced changes in EU

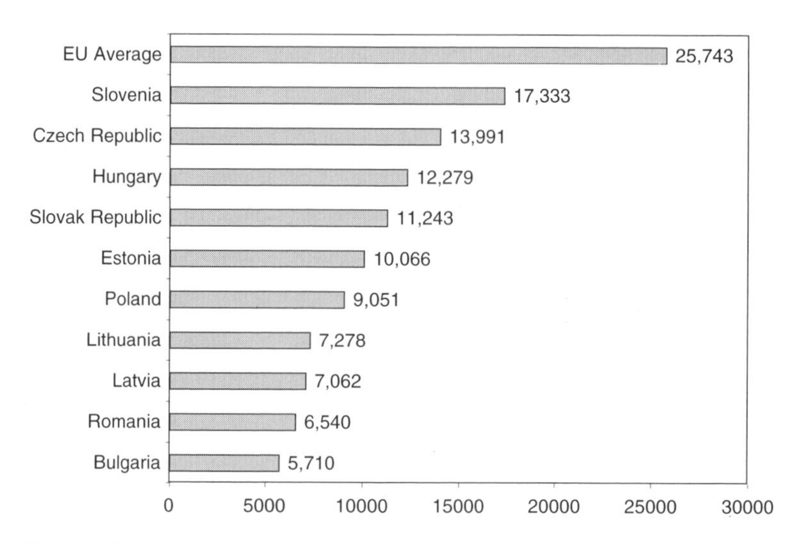

Figure I.2
GDP per capita of Central and East European countries and EU average (2000, purchasing power parity current US$). Source: World Bank 2002. For "Slovak Republic" read Slovakia.

institutions to accommodate larger membership. These changes include modification of the composition of the European Commission, extension of the scope of majority voting, differentiated weighting of countries' votes in the European Council, and new representation rules for the European Parliament. In 2002, the EU decided to admit by 2004 ten new members, including all Central and East European candidates except for Bulgaria and Romania (which aspire to become members in 2007).

Central and East European countries, which have considerably less bargaining leverage than the EU in enlargement decisions, have often evoked normative arguments of democratic cohesion in Europe to speed the accession process.[8] A lot of effort is also placed on the painstaking adoption of the vast body of EU regulations to meet the formal conditions of membership. Thus, while the prospect of EU membership brings long-term security and economic benefits for Central and Eastern Europe, it is also associated with difficult policy adjustment. David Cameroon has outlined most succinctly the multiple challenges of accession: weak ability to administer EU regulations, the high cost of their implementation, the need to proceed with difficult structural reforms and the application of EU competition policies, persistent trade deficits with the EU, high unemployment, and growing "Euroskepticism" among the populations of some candidate countries.[9]

The costs and policy challenges associated with EU accession give rise to concerns about the ability of new members to practically implement EU regulations and cope with the pressures of accession.[10] Such concerns are frequently voiced with respect to the application of EU environment regulations in particular.[11] This body of legislation consists of more than 250 directives and regulations covering general environmental policy, air pollution, water management, waste management, nature and biodiversity, industrial risk, chemical safety, biotechnology, and noise pollution. It is one of the costliest parts of the EU Acquis, estimated to require an investment of approximately 120 billion euros in the accession countries, or an annual investment of approximately 10 million euros over 10 years.[12] The wholesale application of EU environmental legislation presents both administrative and financial challenges for transition countries, which inherited significant environmental damages from more than 40 years of communist economic planning and industrialization. As a consequence, the environment was identified as one of the most difficult areas for accession negotiations by the European Commission's 1997 "road map" document on enlargement, *Agenda 2000*, as well as by international financial institutions.[13]

To facilitate the application of the environmental Acquis, the EU extended assistance for administrative capacity building to accession candidates through its Poland Hungary Aid for the Reconstruction of the Economy (PHARE) program.[14] Investment support for the environment has been provided since the year 2000 under the Structural Instrument for Pre-Accession (ISPA), which covers transportation and environmental infrastructure projects. In the period 1990–2000, the PHARE program allocated 1.06 billion euros and disbursed 593.8 million euros in Central and Eastern Europe for environment and nuclear safety capacity building (table I.3). ISPA allocates approximately 1 billion euros per year for transport and environmental infrastructure in accession states.[15]

EU conditionality and assistance has strengthened environmental protection in post-communist Europe, having helped keep environ-

Table I.3
PHARE assistance to Central and East European countries (1990–2000, million euros). Source: PHARE Program 2000.

Sector	Commitments	Payments
Administration, public institutions	1,265.19	547.00
Agriculture	690.19	488.28
Approximation legislation	141.47	92.35
Civil society, democratization	179.26	86.70
Consumer protection	12.28	11.25
Education, training, research	1,262.45	1,101.51
Environment, nuclear safety	1,056.11	593.81
Financial sector	280.90	259.28
Humanitarian, food, critical aid	598.31	573.63
Infrastructure (energy, transport, telecom)	2,855.44	1,664.44
Integrated regional measures	343.60	193.05
Private sector	1,389.71	994.21
Public health	113.39	98.72
Social development and employment	559.83	249.52
Other	1,132.57	693.51
Total	11,880.70	7,647.26

mental reforms in place despite the decline in political support for environmental objectives during the economic transition. At the same time, the dominance of EU environmental law harmonization is criticized for crowding out the environmental agenda in transition states, diverting attention from domestic priorities, and sometimes preventing the use of policy instruments that are less costly and more appropriate for local conditions.[16] By 2002, despite the anticipated difficulties of negotiating the environmental Acquis, most accession countries had closed negotiations on the environment, with the exception of Bulgaria and Romania, which lagged behind in the accession process overall. Only a few transition periods for compliance were granted to countries that closed environmental negotiations, mostly in areas with significant investment implications such as water treatment, waste management, sulfur content of certain liquid fuels, emissions from large combustion plants, volatile organic compound emissions, and integrated pollution prevention and control.[17]

The relatively high formal adoption of EU environmental regulations raises a number of questions and puzzles. What is the actual implementation of EU environmental norms? Is the environmental area a clear instance of "paper approximation" without implementation, given the high cost of compliance? Although a number of authors have speculated that this may be the case,[18] this book demonstrates that the picture is much more complex and varied. The cases presented here reveal, for example, that Bulgaria, the Czech Republic, and Poland have moved rapidly toward adoption and application of EU chemical safety regulations, despite the cost to industry and the lack of compatible legislation before accession preparations began. In the case of air emissions from large combustion facilities, the Czech Republic, surprisingly, applied EU standards faster than most current EU members; Bulgaria has achieved so far only formal approximation, which is not substantiated with actual implementation. Poland followed a middle path of gradual but consistent application of European air emission standards.

This book posits that in order to understand the patterns of adjustment to EU rules and to gain a better, not just speculative, picture of the effect of EU integration on environmental policies in Central and Eastern Europe, it is necessary to move beyond the inter-governmental level of analysis. We need to examine systematically how international markets and institutions reshape the domestic politics of the environment, and what political processes and policy outcomes ensue. The following section presents a theoretical framework that specifies the role of economic interests, international commitments, and domestic structures in linking international and domestic environmental politics in Europe.

Transnational Politics of the Environment: Theoretical Framework

This book builds on and contributes to two broad fields in international relations theory: EU integration and the internationalization of domestic politics. In subject matter, this study pertains to the vast literature on the EU. Most studies of the EU, however, examine why and how the EU evolved as a system of integration and governance, advancing neo-functionalist, institutional, and inter-governmental explanations.[19] Only recently have scholars, mostly in Europe, engaged the reverse question of EU influence on national policy making, or the "Europeanization" of domestic politics.[20] The Europeanization literature has adopted largely a top-down approach to studying the domestic effect of integration, highlighting how EU institutions and policies constrain and reshape domestic politics, what type of domestic institutions facilitate or obstruct change, and whether patterns of policy convergence or divergence result.[21] Although the institutionalist insights of the Europeanization literature are valuable, they are insufficient for the present analysis. European integration involves not only institutions but also markets. To understand its effects, it is necessary to focus not only on top-down institutional constraints, but also on the bottom-up, transnational influences emanating from market integration that often induce the most dynamic changes in domestic interests and coalitions.[22] This insight was highlighted by early neofunctionalist theories that sought to explain European integration,[23] but so far it has not been applied systematically to the study of its domestic effects.

To illuminate the transnational impacts of European integration on environmental politics in Central and East European states, the analysis draws on political economy theories of policy adjustment under international economic pressures.[24] One part of this literature applies trade theories to understand how international trade redistributes domestic resources and reshapes interests and coalitions, with relevant implications for policy choice.[25] Another part of the open-economy literature, similar to recent studies of Europeanization, highlights the role of domestic institutions in shaping countries' adjustment strategies.[26] Open-economy analyses have been applied mostly to the study of trade and social policies. So far, no compatible framework exists for understanding the impact of economic integration on domestic environmental interests, politics, and regulations.

The literature on international environmental relations was preoccupied for a long time with the development and effectiveness of environmental regimes.[27] Comparative studies of environmental regulations have focused primarily on domestic variables, even in Central and East European states where integration has had such a strong influence.[28] The international-domestic divide in environmental studies has recently been challenged by works that highlight the role of transnational actors (advocacy organizations, epistemic communities, multinational corporations, policy entrepreneurs) in linking the two realms and translating international environmental norms across borders.[29] I contribute to this important trend by extending the open-economy framework to examine the impact of international markets and norms on domestic environmental interests and the role of domestic institutions in shaping environmental policy responses to international pressure. The theoretical analysis that captures these processes first specifies the influence of EU integration on the environmental interests of industrial actors in Central and Eastern Europe as a channel of transnational influence; it then examines the role of domestic institutions in internalizing EU environmental rules.

Integration, Industrial Interests, and Environmental Regulations

There are important reasons to start the analysis of EU influence on environmental protection in Central and Eastern Europe by focusing on the reaction of domestic industrial actors to EU regulations. In regulatory politics, the interests and strategies of industrial groups are important determinants of policy outcomes.[30] Industrial interests also link the international and domestic realms of environmental politics, as they are affected by international trade and investment and have special incentives to ensure that domestic and international environmental standards do not diminish their international competitiveness. Moreover, since industry ultimately bears the implementation cost of many environmental policies, understanding industrial environmental strategies is critical for assessing the level of compliance and effectiveness of environmental standards.

The process of EU integration changes the environmental interests of industrial actors and thus the domestic political dynamics of environmental regulation in Central and Eastern Europe. Industrial environmental interests are defined here to reflect both the environmental preferences of industrial actors (i.e. their ranking of all possible regulatory outcomes on the basis of their perceived costs and benefits) and the strategies of industries in environmental politics (i.e., the political tools used to get as close as possible to preferred policy outcomes).[31] EU integration, I will show in this book, affects the perceived costs and benefits of environmental regulation, as well as the strategic options of industrial actors to influence regulation through three related mechanisms:

international markets incentives, transnational organizations, and governmental commitments to international rules. These effects differ across sectors and firms, reshaping domestic coalitions and the politics of the environment.

International Market Incentives

The linkage between international markets and environmental norms within the EU is one of the main mechanisms of transnational influence on the interest-group politics of the environment in Central and Eastern Europe. In a closed economy, industrial preferences for environmental regulation are determined solely on the basis of the immediate costs and benefits of regulatory compliance.[32] But with access to EU markets linked to a set of environmental standards, business groups in transition countries add to their environmental calculus the gains and losses from free trade.

As theories of trade and political economy have illuminated, international trade redistributes resources domestically, resulting in predictable cleavages between winning and losing firms, sectors, and classes.[33] In Central and Eastern Europe, a pattern of winning and losing industries and sectors of society has also taken shape as a result of regional integration. Regional trade has benefited export-oriented and competitive enterprises, as well as sectors that have restructured and attracted foreign capital such as textiles, certain classes of chemicals, food processing, beverage, furniture and wood processing, automobile assembly, and others.[34] EU integration is also expected to be beneficial for the labor force in transition countries, which is relatively well educated, inexpensive and abundant.[35] By contrast, firms and sectors that produce primarily for the domestic market or compete with imports stand to gain little or even to lose from integration.[36] Andras Inotai, who has extensively studied patterns of East-West trade and efforts at restructuring in post-communist economies, identifies state-owned monopolies in service sectors such as energy, public utilities, railways, and aircraft as substantial losers from integration, unless they are restructured and privatized with the participation of foreign capital.[37]

Many observers point out that these patterns of integration winners and losers may apply for the short and the medium term, as the long-term comparative advantage of firms and entire sectors is likely to be revealed only after the full restructuring of the post-communist economies and the removal of communist-era distortions.[38] For this reason, I will refer to the short- and medium-term preferences of industries when discussing the relationship between markets and environmental interests. I will focus primarily on the division of trade and environmental interests across industrial firms and sectors, rather than on the positions of classes with respect to integration and environmental regulations. In environmental politics in Central and Eastern Europe, interests split along class lines only rarely; workers concerned with jobs and the competitiveness of their sectors often align with industry owners against regulations that would increase the cost of production. The differential costs and benefits that result in Central and Eastern European societies from regional integration have important implications about how industries react to environmental requirements associated with EU integration. Economic actors that gain from integration, namely exporters and multinational firms, have distinctive incentives to support EU enlargement and the associated requirements as long as the costs of regulation do not outweigh the benefits of integration.

For exporters, the adoption of EU standards can improve access to the EU market. Compliance with EU standards for products and even for processes is often a precondition for subcontracting to EU firms. Harmonization also removes potential barriers to trade for export-oriented firms and wards off accusations of ecological dumping. Moreover, the adoption of EU norms may even open new "green niche" markets and opportunities for product differentiation in environmentally sensitive markets, providing commercial incentives to support the "ratcheting up" of domestic standards to those of large regulated markets.[39] As one industry representative remarked in an interview, "the environmental codes associated with international markets make the consideration of environmental performance a necessity for East European firms involved in international transactions."[40]

Of course, exporters could also improve their access to EU markets and their market share within the EU by seeking to avoid costs associated with environmental abatement and by offering lower prices. This may be particularly attractive for small enterprises, for which the relative cost of regulation is higher.[41] However, in many cases the practical success of such a strategy is limited by consumer preferences, by subcontracting requirements, or by pressure from industry and consumer organizations. Moreover, full EU membership, which is generally supported by export-oriented firms and sectors since it removes the threat of reversal in free trade for sensitive sectors and ensures long-term economic stability, is contingent on regulatory harmonization. This provides an additional incentive to forward-looking firms with international interests to incorporate EU standards early on and to seek policy mechanisms that minimize the implementation costs.

The adoption of EU standards also reduces transaction costs for exporting firms. This is especially true for large firms with extensive international operations. The position of chemical and other large export-oriented enterprises described in the empirical chapters of the book reveals, for example, that industries involved in international transactions quickly develop a preference for stable and uniform regulations associated with EU accession over weaker but more uncertain systems.[42]

Finally, the harmonization of EU standards gives exporters an advantage over domestic competitors because benefits of harmonization associated with easier trade incur disproportionately to internationally oriented firms and increase the relative cost of domestic production. Exporters also tend to have greater capacity for compliance with international standards than domestic firms, insofar as they benefit from economies of scale and from more ready access to cleaner technology and knowledge of international standards. The costs associated with stricter standards increase the entry barriers for new producers, further enhancing the benefits of firms with established international positions.

Thus, the prospect of easier access to the EU markets, reduced costs of international transactions, a stable regulatory environment, and advantages over domestic competitors provides distinct commercial incentives for Central and East European exporters to support the adoption of EU environmental standards. These incentives are stronger for large export-oriented firms than for small export-oriented enterprises, as the latter do not benefit from economies of scale and bear higher relative costs of environmental regulation. Similarly, EU integration is likely to provide weak environmental incentives for exporters that might target exclusively non-EU markets such as Asia, the countries of the former Soviet Union, and the Americas. However, the predominance of the EU in the trade structure of Central and Eastern Europe and the dominance of large firms in shaping the political strategies of sectors imply that the overall pull of EU markets and environmental rules is likely to be strong, creating a distinctive political constituency that supports harmonization.

Multinational enterprises, which are also among the beneficiaries and supporters of EU enlargement, have similarly distinct incentives to support the adoption of EU environmental standards. Multinationals, especially those from EU countries, worry less than Central and East European exporters about access to EU markets, which is typically ensured by their already-established production networks and marketing strategies. However, multinationals are under increasing scrutiny from consumer unions, advocacy groups, and even shareholders to apply in host countries environmental standards compatible with those in home countries. Moreover, similarly to exporters, multinational enterprises reap benefits from harmonization in terms of reduced transaction cost of operation, greater regulatory stability, and advantages over domestic firms and new entrants. Multinationals, even more so than export-oriented firms, take advantage of easier access to technology and management skills to apply international standards at a lower cost and even to market their own cleaner technologies in Central and Eastern Europe. Interestingly, even non-EU firms that invest in Central and Eastern Europe tend to support regulatory harmonization with the EU, as they value regulatory stability and uniformity across Europe more than freedom to bargain with local authorities for weak environmental regulation.[43]

In sum, the high level of market integration between Western and Eastern Europe, the linkage between markets and environmental regulations within the EU, and the prospect of EU accession alter the perceived costs and benefits of adopting EU environmental regulations for important segments of industry in accession states, namely firms and sectors that are export-oriented and multinational. Such firms and sectors are likely to support the adoption of EU standards. By contrast, for firms and sectors that lose or do not benefit from integration, namely those that compete with imports or supply the domestic market, EU accession does not provide any special incentives to upgrade environmental performance. On the contrary, the extra cost of regulation may compound the disadvantage of domestic firms and sectors and may provide an additional rationale for blocking harmonization.

The Role of Transnational Business Networks

EU integration influences the environmental position of internationally oriented industries in Eastern Europe not only through the invisible hand of the market but also through pressure and assistance from transnational business organizations. As functionalist theories anticipated in the 1950s and the 1960s,[44] European integration has facilitated the emergence of pan-European economic interests and organizations that link business entities across borders and represent them in EU institutions.[45] In international environmental politics, business is also increasingly organized transnationally to promote a set of market-based environmental ideologies and to influence global environmental affairs.[46]

Transnational business organizations and networks have influenced the environmental strategies of internationally oriented businesses in Central and Eastern Europe in several ways. First, EU business associations such as the European Chemical Industry Council (CEFIC) provide information about relevant EU standards, increasing the environmental sensitivity of East European exporters and their awareness that EU standards can be used as barriers to trade. Driven by motivation to avoid competitive disadvantage, EU business organizations also exert direct pressure for the adoption of EU standards, making it clear that their support of EU integration is contingent on compliance with EU norms.[47] Transnational organizations also assist businesses in transition countries with information, training, and the promotion of clean technologies, reinforcing market pressure for improved environmental performance.[48]

Domestic industry associations are the main interlocutors of transnational organizations and the main beneficiaries of their assistance. Industrial sectors in Central and Eastern Europe inherited well-organized sector and peak associations from their communist past, which before the democratic changes of the late 1980s were largely subservient to the state. During the transition period, sector and other business associations sought to reinvent themselves as lobbyists for industrial interests. In this process, cooperation with European and other transnational organizations provided valuable political and organizational resources. Capacity building by international business networks has stimulated the establishment in Central and Eastern Europe of a new type of "green business organization," intended to diffuse the environmental stewardship ideology and increase the influence of business over domestic environmental policy.[49]

This dynamic is illustrated in detail by the chemical industry cases presented below in chapter 1, where transnational assistance made national chemical associations central players in the adoption of EU chemical safety legislation and its domestic implementation. While chemical associations typically gave weight to the interests of large export-oriented members, they also sought to establish compensation mechanisms for smaller firms and to strengthen the overall support for integration. Intra-industry compensation schemes include free training and subsidized auditing and consulting services for smaller firms and, in some cases, lobbying for a more gradual schedule of compliance for small and medium-size enterprises. Transnational and domestic associations thus played an important role in resolving problems of collective action among industrial actors and in shaping a pro-integration and pro-harmonization lobby in the environmental politics of accession states.

In summary, both through market incentives and through organizational incentives, EU integration has influenced industrial environmental interests in Central and Eastern Europe, reshaping the politics of the environment from the bottom up. The argument yields a number of observable implications. It indicates that, even in countries with different histories and different policy-making traditions, export-competitive sectors and export-oriented and multinational firms should follow similar strategies, supporting both EU integration and the adoption of EU standards that facilitate trade. The observation of similar reshaping of domestic interests and coalitions across accession states would therefore be a strong indication that EU markets and rules do have a systematic influence over domestic politics. Moreover, if the environmental influence of European integration depends on the economic characteristics of actors, changes in these characteristics should be associated with reevaluation of their environmental positions. For example, if, in the process of restructuring, previously publicly owned sectors or firms attract significant international investments, we should expect their environmental strategies with respect to EU regulations to change in a positive direction. By tracing these observable implications and the changes in the environmental preferences and strategies of the chemical and electricity industries in three accession countries, this book uncovers powerful transnational sources of EU influence on domestic environmental politics that typically escape the purview of state-centered analyses.

Government Commitments and Domestic Interests

In addition to the bottom-up effect of integration on domestic environmental interests, the EU influences environmental policies in accession states through government commitments to adopt the Acquis Communautaire as a condition for membership. This is the most widely recognized mechanism of policy influence in accession countries, and it is a classical instrument of international influence on state policies.[50] Most accounts of EU enlargement also present the adoption of EU rules by accession countries as driven almost exclusively from above by empowered executives and the European Commission, with limited involvement of societal interests and parliaments.[51] This book offers a considerably more complex picture of domestic adjustment to EU conditionality—a picture in which government commitments interplay with domestic and transnational interests and institutions to shape policies.

The change in the environmental interests of export-oriented and multinational industrial actors as a result of integration has strong implications, for example, about the relative ease with which governments are able to adopt EU standards. The interest-based analysis implies that the presence of a new constituency supportive of EU harmonization is likely to result in higher overall levels of EU environmental harmonization than might be anticipated if only international commitments and the overall cost of compliance are considered. It also implies that domestic incentives for harmonization and compliance will be especially strong for environmental regulations that pertain to the common European market or affect the activities of export-oriented and multinational sectors, such as auto emission standards and lead content in fuels affecting the highly multinational automobile sector, product and process standards affecting the internationally oriented chemical industry, or safety and environmental standards targeted exclusively at the food processing industry. In such areas of regulation, domestic interests, government objectives, and EU norms are closely aligned. As a result, the tendency will be for a high level of compliance with EU standards by industry and regulatory convergence driven by transnational interests and international norms, despite institutional differences among countries. This logic is presented in the left branch of figure I.3, which illustrates the mechanisms of change in domestic environmental politics under the influence of EU markets and institutions.

The adjustment of national environmental policies to EU standards is not so unidirectional, however, in cases when EU regulations affect multiple sectors having different positions with respect to enlargement and impose costs on actors that do not benefit from integration. In a host of regulatory areas, such as emissions from large combustion facilities, water treatment, waste treatment, and sulfur content of fuels, EU regulations create concentrated costs for private and public industries and utilities, which service largely domestic users and which do not enjoy offsetting benefits of international integration. In such instances, EU commitments still affect the strategies of domestic actors subject to regulation, as these actors respond to international pressure for policy reforms either by seeking to block change or to gain compensation and extended time periods for compliance. Under such a constellation of interests, therefore, harmonization will be more difficult to achieve and more variable across states. The success of harmonization efforts depends more closely on governments' ability to overcome particularistic opposition and the presence of what is labeled as "enabling" domestic institutions in the right branch of figure I.3, resulting in variable level of adoption and compliance with EU regulations across states. This dynamic also suggests that in order to gain further insight into the variety of policy responses to European integration we need to look deeper into domestic politics and unpack the notion of "enabling institutions."

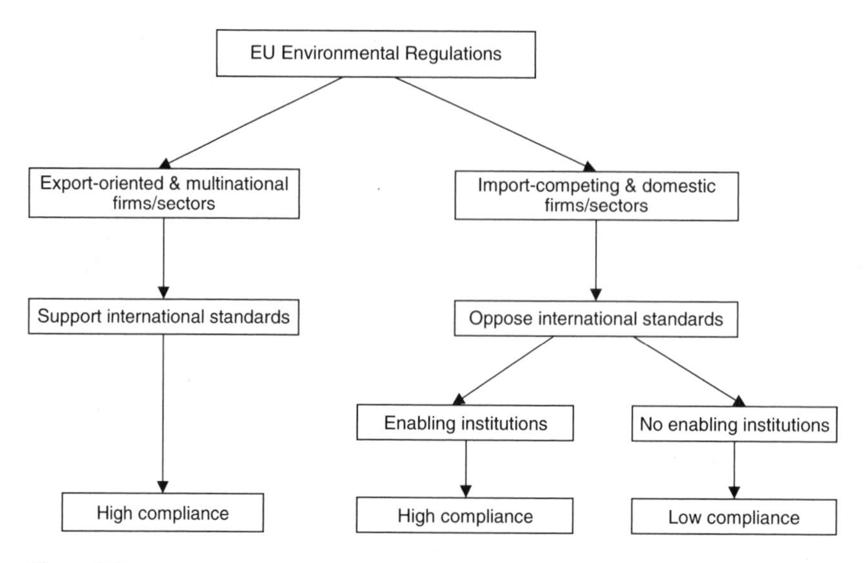

Figure I.3
EU influence on environmental politics in Central and Eastern Europe.

Integration, Domestic Institutions, and Policy Outcomes

Studies of the "Europeanization" of domestic policies of EU member states and institutional approaches to analyzing external economic influences on national policies converge in their emphasis on the role of domestic institutions in mediating between the international and domestic realms.[52] Studies in these traditions highlight the roles of institutional factors such as veto players,[53] institutions for collective bargaining and compensation,[54] and the structure of the political process[55] in shaping divergent national policies under similar international constraints. To illuminate the influence of integration on environmental policies in Central and Eastern Europe, I integrate the insights of the institutional and interest-based perspectives to ask what type of domestic institutions might facilitate compliance with EU standards when they impose high domestic costs without significant offsetting benefits of integration. Three institutional characteristics appear particularly pertinent to this analysis: the veto position of regulated actors, the capacity for interest mediation and compensation, and the strength of environmental movements.

The Veto Position of Regulated Actors

Studies of EU Acquis implementation by member states and of policy reform more broadly identify the number of veto points in the decision-making process as a critical variable accounting for policy transformation or lack thereof.[56] The empirical analysis in this book also puts a strong emphasis on identifying the veto position of relevant actors in environmental politics, particularly in instances when the process of integration does not create interests supportive of reform. The concept of veto power is specified here more broadly to reflect not merely the number of structural veto points but also other institutional characteristics that accord a blocking position to political actors.

For example, even if the policy-making processes of two countries are characterized by a similar number of formal veto points, it is important to know whether veto actors are strongly responsive to a particular interest group or whether they represent the interests of a broader constituency. Both improved environmental quality and EU membership are likely to increase the general welfare of post-communist societies while imposing particularistic costs. If a veto actor is accountable to a narrow constituency, it is more likely to block regulation that is welfare enhancing for the country but imposes concentrated costs for specific groups than a veto actor with a broad constituency.[57] It matters whether the veto actor is the presidency (with a national constituency) or a sectoral ministry (which represents the interests of particular sectors).

The veto position of societal interests can also be determined by interest representation in pivotal political organizations. A party in a coalition government with a narrow majority will have a pivotal position and a considerable veto over legislative proposals regardless of the party's size and electoral base. If such party represents a narrow societal group (for example, an ethnic minority, agriculture, or environmentalism), the interests of this constituency will have a disproportionate influence and a de facto veto over policies. A strong labor union with influence over party politics can similarly be in a position to veto regulatory changes.

The Central and East European countries examined in this book differ along some of these dimensions in important ways. In two of the countries, Poland and Bulgaria, air pollution regulations are specified by executive ordinances on the basis of a broad framework legislation. In the Czech Republic, air pollution norms are elaborated more fully in comprehensive sector-specific laws. Under the former regulatory system, governmental ministries have a stronger veto over the content of regulations; under the latter, greater veto power is concentrated in the hands of parliaments, which tend to have broader constituencies than sectoral ministries. In addition, of the three countries, Poland has the strongest labor unions, with direct political influence over both left-wing and right-wing parties and the ability to block policies that will hurt highly unionized sectors such as coal, power, and steel production. The case studies use the concept of veto power to understand the different structure of the political game in each country and the ways it shapes the adjustment of environmental policy to costly international rules.

Institutional Capacity for Compensation

When international pressures create uneven costs and risks of adjustment, the ability to compensate losers or to establish acceptable societal bargains can be crucial for the success of national adjustment strategies. This insight, first underscored by studies that examined how industrialized societies responded to international economic shocks, has more recently been reemphasized by the literature on globalization.[58] The political economy literature highlights the role of such specific compensatory arrangements as corporatist political structures,[59] the coexistence of strong left-wing parties and encompassing labor movements,[60] and increased government spending to compensate for insecurities associated with global integration and volatility.[61]

But what is the significance of redistributive mechanisms in facilitating national compliance with international environmental norms? This question is rarely engaged directly in the environmental policy literature, partly because of the dominance of the "polluter pays" principle, which assumes that polluting entities will not be compensated by society for reducing environmental harms. Yet, while the "polluter pays" principle guides environmental policy, mechanisms that streamline or redistribute the cost of compliance with domestic and international environmental regulation (such as grandfather clauses and extended implementation schedules) have been used widely in industrialized countries to overcome policy deadlock.[62] Studies of the effectiveness of international environmental regimes also emphasize the roles of domestic capacity and international assistance in compensating the cost of international agreements and facilitating their implementation.[63]

In countries undergoing economic transition, institutional capacity and policies that offset the incremental costs of environmental improvements have been of tremendous significance for environmental reform and compliance with international standards. This is not surprising. The more interesting question that this book illuminates is what kind of compensatory mechanisms have been successfully used and what determines the different ability of countries to apply such instruments. The cases presented demonstrate that, although economic development certainly creates greater capacity for compensation, economic growth is not the only condition for establishing compensatory bargains. Mechanisms that alleviate the cost of environmental improvements range from incremental subsidies for environmental infrastructure to tax incentives, preferential financing, cost-minimizing economic instruments of regulation, phased-in compliance periods, subsidized research and knowledge creation, and the promotion of win-win technological solutions.

The national funds for environmental protection are an example of an institutional form established throughout transition states for targeted environmental financing for the environment.[64] But funds across Central and Eastern Europe differ both in capacity and in ability to influence environmental policy and instruments. Table I.4 provides comparative figures on the resources of the Bulgarian, Czech, and Polish environmental funds. The Polish system of central, regional, and local funds is not only richly endowed; it is also a politically influential system of institutions as a result of their strong links to local politics and regulators. Poland also pioneered the establishment of other mechanisms of "green financing" through its Environmental Protection Bank, a debt-for-nature swap, tax relief, and emphasis on economic incentives in regulation.[65] As a result, Poland, which is relatively poor compared to EU countries (figure I.2) although more advanced economically in comparison with most transition countries, has created conditions for stimulating environmental investments comparable to (as share of GDP) if not greater than those in the EU. The air pollution case studies show that, as a result, the environmental

Table I.4
Resources of national environmental funds in Bulgaria, the Czech Republic, and Poland, 1993. Source: Vukina et al. 1999, p. 588.

	Million 1987 US$	Percentage of GDP
Bulgaria	1.8	0.01
Czech Republic	84.1	0.28
Poland	156.1	0.27

administration of Poland was able to use a whole artillery of incentives to gradually renegotiate the environmental position of the powerful electricity sector, which at the beginning of the 1990s looked set to resist any introduction of European air emission standards.

Bulgaria, with the weakest capacity for stimulating environmental investments, is an instructive example at the other end of the spectrum. This is a consequence not only of its relatively low national income, but also of delays in structural and institutional reforms. The environmental fund of Bulgaria remained not only poorly endowed but also poorly managed throughout the 1990s; it was eventually closed in 2002 under pressures to streamline the national budget. Other mechanisms of environmental financing have been weakly developed or non-existent in the country. As a result, the ability and willingness of industrial enterprises and public utilities to invest in environmental improvements have been minimal despite a strong national commitment to the adoption of EU standards.

The Role of Environmental Movements

The organizational and political strength of domestic environmental movements is another important factor for the success of environmental reforms and the implementation of international commitments.[66] In post-communist Europe, where public concern for the environment dropped quickly during the transition period, I expected that environmental groups would seek to exploit the salient issue of EU integration to advance their agendas. Interestingly, it turned out that, for a range of political and ideological reasons, environmental groups rarely evoked EU conditionality.

The relatively weak involvement of environmental groups in EU environmental law approximation reflected at first their limited information about the nature of EU regulations and the harmonization process.[67] Only in the second half of the 1990s did EU-based environmental organizations, such as Friends of the Earth Brussels and Milieukontakt Oosteuropa, begin to sponsor projects to familiarize their Eastern European counterparts with the environmental Acquis and to increase their voice in the accession process. In addition, the priorities of post-communist environmental groups have been determined largely by international donor and conservation organizations and emphasized biodiversity, access to information, and accountability of multilateral agencies.[68] There has been considerably less attention paid to other issues regulated by EU law, such as water and air contamination and even industrial pollution—precisely some of the areas that are the focus of this study.

Environmental groups in Central and Eastern Europe have been profoundly ambivalent about the environmental consequences of EU enlargement. On one hand, advocates recognize that EU conditionality helps to sustain reforms; on the other hand, they are concerned about the environmental "side effects" of integration associated with the sprawl of highway transportation, with growing consumerism, and with the subsidization of agriculture.[69] East European advocates have also opposed waste incineration, a method of waste management endorsed by the EU. They have pointed out that in some cases EU standards replace stricter (albeit not enforced) domestic requirements and tend to impose end-of-pipe solutions rather than efficiency-enhancing options.

The EU Environmental Impact Assessment regulation is probably the area of EU conditionality evoked most readily by environmentalists in Central and Eastern Europe to promote their fundamental objective of easier access to information. But even in this area broader transnational cooperation among European advocacy organizations, which culminated in the adoption of the pan-European Aarhus Convention on Access to Information (1998), has dominated the strategies of environmental groups more than EU enlargement.

As a result of the limited role of environmental organizations in the EU accession process, some important differences in the organization of the Bulgarian, Czech, and Polish movements did not have strong differentiation effects on how each country adjusted to EU air pollution and chemical safety directives. For example, the broad-based environmental movement in Poland, which has strong central organizations as well as regional and local networks, did not get actively involved in the reform of chemical safety regulations, similarly to its Czech and Bulgarian counterparts. Even in the area of air pollution, in which a number of Polish groups have well-developed activities, the policy process in all three countries was dominated by governmental agencies, industrial actors, and parliaments without substantial input from advocacy organizations.

Cross-country differences in the position of environmental movements influenced environmental policies most significantly in the early post-communist period. The strong anti-nuclear Polish movement, for example, halted the development of nuclear plants in the aftermath of the democratic transition, constraining the air pollution mitigation and bargaining options for the power industry later on. The presence in the first post-communist Czechoslovak government of prominent dissident environmentalists, who pushed for a strong Act on Clean Air, similarly shaped the strategic options for bargaining between government regulators and the electricity industry during the 1990s. Beyond this early formative influence on certain policies and institutions, however, the environmental movement has played a relatively small role in the process of EU environmental law harmonization beyond the areas of its immediate interests, such as access to information and biodiversity conservation.

Research Design and Summary of Findings

This book offers a two-tier approach to understanding the interaction between international and domestic environmental politics in Central and Eastern Europe. It posits that EU integration has altered the scope of domestic support for environmental reforms, creating strong incentives among export-oriented and multinational sectors to support integration and regulatory harmonization. Where such incentives are absent, however, the structure and capacity of domestic institutions play a strong role in facilitating or impeding the internalization of EU standards. The argument integrates the insights of interest-based and institutional theories to shed light on the profound transformation of domestic environmental politics in Central and Eastern Europe as a consequence of integration with the EU.

To evaluate the explanatory power of this perspective, comparative case studies are used. In-depth case analysis is a method well suited for uncovering the process of interaction between regional and domestic politics, rather than solely explaining outcomes.[70] This type of research design draws on contemporary material to document the change (or lack thereof) in the environmental interests and strategies of domestic groups, the influence of these changes on national policies, and the significance of domestic institutions in mediating between international pressures and environmental regulations. The case study method also provides an opportunity to document the extent to which the empirical evidence matches the observable implication of the theoretical argument, offering more robust support for the theoretical claims.

The cases selected for this project examine the effect of EU markets and institutions on the development of chemical safety and air pollution policies in Bulgaria, the Czech Republic, and Poland, and the position of the chemical and electricity industries in this process. There are two dependent variables: (1) environmental policy preferences and political strategies of the regulated industry and (2) environmental policy reform and harmonization with EU environmental legislation.

I infer changes in the environmental preferences of industrial actors from statements of anticipated costs and benefits associated with environmental regulations and European integration. I also examine statements of the preferred course of regulatory reforms presented in industry documents, memos, and interviews. In addition, I document the strategies of industrial actors and societal groups on the basis of their choice of political allies, parliamentary activities, and interaction with the government.

The reform of environmental policies, the second dependent variable in the study, is measured in two steps. First, the cases examine the formal changes in national legislation or regulations during the period 1990–2001 and the level of harmonization with EU standards. Second, the analysis evaluates the adoption of mechanisms for implementation and measures for improved environmental performance.

In the research design of the study, I selected two industries that represent a relatively low and a relatively high level of export orientation; I examined a set of environmental regulations that affect them most directly. Of the range of EU air pollution regulations, the book examines the adoption of the Large Combustion Plant Directive and related regulations under the Long Range Transboundary Air Pollution (LRTAP) convention that seek to limit air emissions from combustion utilities. These international standards primarily affect the electricity sector, which produces largely for the domestic market. In the period 1990–2000, between 2 percent and 10 percent of the electricity generated in Bulgaria, the Czech Republic, and Poland was exported, primarily to non-EU markets. By contrast, chemical production, which is the main target of EU chemical safety regulations, is a highly export-oriented and increasingly multinational sector in each of the three states examined. According to data provided by the UN Statistical Division, chemical products are among the top ten commodities exported by Bulgaria, the Czech Republic, and Poland.[71] This selection of industries makes it possible to evaluate the proposition of the differential influence of EU markets and environmental norms on sectors characterized with different degrees of international exposure and competitiveness.

The selection of Bulgaria, Poland and the Czech Republic was motivated by some important similarities of their chemical and electricity sectors and the environmental problems associated with their activities. In all three countries the chemical sector is a significant branch of industry, strongly dependent on exports and relatively attractive to foreign investors. This makes it possible to trace whether there is a similar process of an externally driven adjustment in the environmental strategy of the chemical industries in the three countries despite economic and institutional differences.

In the cases of air emissions from electricity utilities, Bulgaria, the Czech Republic, and Poland inherited from their communist past similarly high levels of pollution due to lack of pollution-abatement equipment and their strong reliance on indigenous brown coal for fuel, which implies high costs to meet European air emission standards in the three states. The ability to control for fundamental factors such as levels of air pollution inherited from communism and the cost implications of European standards enables the air pollution case studies to isolate more clearly the role of domestic institutions and capacity in facilitating the adoption of international environmental standards that impose concentrated domestic costs.

The selection of countries has the additional advantage of comparing the experience of two front-runners in terms of economic reforms and preparedness for the EU accession (the Czech Republic and Poland) with a country that has lagged in economic transformation and is in the second tier of EU applicants (Bulgaria). Such comparisons are rarely attempted in post-communist studies; scholars have focused mostly on single-country cases or on comparisons within the leading or the lagging group. The more encompassing inquiry presented here makes it possible to see the extent to which the commitment to EU integration causes similar hurdles and domestic political reactions across all candidates, and also to see the relative importance of economic and institutional capacity for tackling accession challenges.

In researching the cases of chemical safety and air pollution regulations in Central and Eastern Europe, I drew on extensive knowledge of the region and on the ability to use primary documents. The fieldwork in Bulgaria, the Czech Republic, Poland, and at the European Commission involved interviews with government officials, industry representatives, environmental groups, and EU officials (see appendix), examination of original reports and internal memos of governmental institutions and industry associations, and work with parliamentary records.

The detailed look at the environmental politics of the Czech Republic, Poland, and Bulgaria revealed fascinating evidence of the varying impact of European integration on domestic interests and coalitions. Tables I.5 and I.6 summarize the change in the politics of chemical safety and air pollution regulation in Bulgaria, Poland, and the Czech Republic. In accordance with the interest-based argument advanced in the book, there is evidence of rapid adjustment in the interests of the highly internationalized chemical sector, its participation in transnational business networks, and ultimately in its support for harmonization with

Table I.5
Politics of compliance with EU chemical safety regulations.

	Czech Republic	Poland	Bulgaria
Industry interests	Support adoption of international standards	Support adoption of international standards	Support adoption of international standards
Industry strategies	Proactive position in environmental policy	Proactive position in environmental policy	Proactive position in environmental policy
	Voluntary standards	Voluntary standards	Voluntary standards
	Promote harmonization	Promote harmonization	Promote harmonization
Policy outcome	Rapid harmonization	Rapid harmonization	Rapid harmonization
	Implementation	Implementation	Implementation

Table I.6
Politics of compliance with 1988 EU Large Combustion Plant Directive and 1994 Second Sulfur Protocol to the Long Range Transboundary Air Pollution (LRTAP) convention.

	Czech Republic	Poland	Bulgaria
Industry interests	Oppose adoption of EU standards	Oppose adoption of EU standards	Oppose adoption of EU standards
Industry strategies	Weak resistance	Delay reforms	Block reforms
	Compensation	Study cost of compliance	Resist compliance
	Compliance	Bargain for compensation	
Policy outcome	Early harmonization	Gradual harmonization	Formal harmonization
	Early implementation	Gradual implementation	Weak implementation

EU directives (table I.5). It is interesting to note that institutional differences across countries had a limited impact on national adjustment to EU chemical safety standards, a finding that highlights the role of transnational interests in promoting regulatory convergence in Europe. In the Czech Republic, Poland, and Bulgaria, chemical safety policies have undergone dramatic changes, very much in compliance with EU regulations and with the active support of the chemical industry (table I.5). Where differences in institutional arrangement and capacity hindered reforms, the involvement of international organizations compensated for national deficiencies and contributed to similar policy outcomes.

The cross-national similarity in the reform of chemical safety regulations is even more striking when contrasted to the dissimilar routes taken by Bulgaria, the Czech Republic, and Poland in reforming their air pollution policies. For the electricity sector, EU integration did not provide motivation to support costly air pollution regulations. International standards were used as a commitment mechanism by national governments, but with different degrees of success in the three countries (table I.6). In the Czech Republic the government was able to follow a dual strategy of coercion and compensation to achieve a surprisingly rapid harmonization and implementation of European air pollution standards. In Bulgaria, the realization of the government's commitment to the adoption of EU standards was successfully resisted by the electricity sector. This resulted in a "hollow harmonization" of EU legislation—that is, a harmonization of EU environmental norms in ways that allow for extensive exemptions and little emphasis on implementation. The case of Polish air pollution reforms is an intermediate one. Poland achieved gradual adoption of European standards, despite the strong initial opposition from the power industry, through a domestic strategy of iterated bargaining with the electricity sector backed by considerable soft financing to facilitate investments in pollution-abatement technology. The research presented in this volume thus uncovers a transnational picture of environmental politics in Central Eastern Europe, one in which international and domestic political and economic incentives interplay to determine policy choice. Challenging the long-standing division between comparative and international environmental studies, it illuminates patterns of cooperation and environmental policy making in Central and Eastern Europe that would be difficult to account for by theories that focus solely on international or domestic levels of analysis.

Contribution to International Relations

This study contributes to the international relations literature the first systematic theoretical and empirical analysis of the effect of EU integration on the domestic environmental politics of Central and East European states. It allows the reader to gain a detailed perspective on how this central foreign policy objective of post-communist states alters domestic political dynamics, challenges established interests, and interacts with political institutions to shape policies. Specifying the mechanisms of international influence is important for the scholarship on post-communist transitions and European integration, insofar as it provides a firmer understanding of the challenges and opportunities associated with integration and the influence of the EU on national policies.

Beyond capturing important transformations in European politics, this book adds to the broader political science literature that bridges the traditional divide between international and domestic politics. In an era of growing globalization and regionalism, it is particularly important for both scholars and policy makers to understand the pervasive effects of international forces on domestic groups and politics. This book offers a generalizable framework for tracing the differential impact of international markets, transnational organizations, and norms on domestic interests in regulatory politics and the implications for policy adjustment under international constraints.

The contribution of a two-level analysis of environmental politics is particularly important for environmental studies, which were long split between analyses of international regimes and analyses of domestic regulations. Scholars who examine the influence of transnational actors recently questioned this split. Pressing policy questions about the implications of international markets, rules, and globalization for environmental protection also demand multi-level analysis. The focus of the present study is precisely on the interaction of international and domestic incentives (and disincentives) for environmental protection. It highlights the dual influence of international markets and institutions on the environmental strategies of states and sub-state actors, and it illuminates the conditions for the different ability of states to meet international environmental norms.

Structure of the Book

The book is divided into two parts, reflecting the research design of the study. Part I focuses on the chemical industry and on chemical safety regulations. Part II examines the electricity industry and air pollution regulations. The two parts are structured differently to capture the distinct dynamics of domestic politics adjustment in the two sectors. In part I, chapter 1 examines the influence of EU markets and regulations on the interests of the chemical industries in Bulgaria, the Czech Republic, and Poland, highlighting common, internationally induced dynamics. Chapter 2 follows the reform of chemical safety policies in the three countries.

In part II, chapter 3 examines the environmental position of the electricity industry and the reform of air emission policies in the Czech Republic. Chapters 4 and 5 develop similar case studies for Poland and Bulgaria respectively. The division of the air pollution cases along country lines reflects the fact that the environmental strategy of the electricity industry with respect to European regulations is strongly conditioned by the domestic institutional environment and is, therefore, best described in reference to the policy context of the particular country. The structure of part II thus underlines the scope for divergence of national industry and policy responses to European pressures.

The conclusion summarizes the case studies and the implications of the analysis for policy and international relations theory.

I

Chemical Safety Regulations: The Harmonizing Influence of International Markets and Institutions

The legislation regarding chemical substances includes some of the oldest and most complex parts of the European Union's environmental Acquis. Directive 67/548/EEC, the first of a number of directives aimed at protecting human health and the environment from harmful chemicals, was adopted in 1967. It regulated the classification, packaging and labeling of more than 1,000 dangerous substances. Since the goal of environmental protection was not included in the 1957 Treaty of Rome, which established the European Economic Community, most chemical safety standards were initially based on article 100 of the treaty, which provided for the harmonization of regulations directly related to the functioning of the Common Market.

Regulatory activities to control chemical substances intensified in all industrialized societies during the 1970s and the 1980s, when public concern about the environmental and health impacts of chemicals became one of the precursors of modern environmental movements. The introduction of an array of national chemical safety standards, coupled with the growing internationalization of chemical production and trade, also provided incentives for international cooperation and cross-border harmonization of chemical regulations. The passage of the Toxic Substances Control Act in the United States in 1976 gave strong impetus to the harmonization of chemical safety regulations within the European Community. Large chemical producers and national officials in Europe were concerned that the provisions of the US legislation on chemicals would limit the access of European chemical products to the US market. In response, EC members sought to align their national chemical policies further and to consolidate a common regulatory framework. This motivated the rapid adoption of the sixth amendment to directive 67/548/EEC, which was a major step toward strengthening the chemical safety system in the EC. The harmonization of chemical standards within the Common Market facilitated the development of a broader international system of chemical control in the framework of the Organization for Economic Cooperation and Development (OECD).

The 1977 Seveso accident, which resulted in the contamination of a large area of Northern Italy with dioxin, further increased the societal pressure for safer production, use, and disposal of chemicals within the EC. The authority to introduce environmental legislation was also strengthened by subsequent amendments of the Treaty of Rome. The Single European Act (1987) recognized environmental protection as a legitimate policy goal of European institutions and introduced a series of articles (130r, 130s, 130t, and 100a) on the basis of which community-wide environmental standards could be adopted. The Maastricht Treaty of the EU (1992) further allowed the use of majority voting by the European Council on environmental legislation, facilitating the adoption of new environmental regulations. Over the years, the EC and later the EU established a comprehensive system of laws for the safe production, use, and disposal of hazardous substances.

One important part of the EU chemical legislation covers the classification, labeling, packing, and notification of dangerous substances. This legislation specifies mandatory testing requirements for substances not listed in the 1981 European Inventory of Existing Chemical Substances. The requirements include submission of a set of toxicological studies evaluating risks for health and the environment, specification of harmful effects in possible uses, proposed labeling and classification, and safety data sheets for dangerous substances. Companies marketing substances not listed in the inventory have to undertake a notification process even if another company previously tested and marketed the same substance, but can use test data from earlier notification processes. Chemical safety directives also classify products according to 15 categories of danger and introduce labeling requirements including danger symbols, information on risk and safe use, and in some instances a requirement for safety data sheets.

Directive 88/379/EEC and subsequent amendments introduces provisions for the classification, packaging, and labeling of dangerous preparations (i.e., mixtures or solutions of two or more substances). Regulation EEC 793/93 mandates the evaluation of risk and requires submission of information on toxicity for existing substances, on the basis of which the European Commission identifies certain existing chemicals for immediate attention because of their potential effects on health and the environment. Other directives prohibit or restrict the marketing and use of substances such as asbestos.

The EU chemical legislation also bans certain classes of pesticides (directive 79/117/EEC) and controls the placing of pesticide products on the market (directive 91/414/EEC). In 1998, the EU adopted a directive regulating the safety of biocidal products (directive 98/8/EC). That directive covers disinfectants, chemicals used for preservation of products and materials, non-agricultural pesticides, and anti-fouling products used on hulls of marine vessels. EU laws also establish a system for notification of the import and export of dangerous chemicals to third countries. The directive on Good Laboratory Practice (87/18/EEC) aligns EU requirements for testing and assessing the risk of chemicals with the system adopted by the OECD. The Seveso I and Seveso II directives (82/501/EEC and 96/82/EC, respectively) set procedures to reduce the risk of major industrial accidents and to limit the consequences when such accidents occur.

The EU chemical safety legislation thus establishes a complex set of rules for assessing the risks associated with the use and production of chemicals, providing information on those risks and hazards, limiting their harmful effects, and minimizing the risk of major industrial accidents. Because a compatible system of chemical control was not in place in the post-communist countries, the application of EU chemical legislation by Central and East European candidates implies a considerable administrative burden for the accession states and compliance costs for the chemical industry. The introduction of chemical safety regulations was also not an immediate priority for post-communist governments in the aftermath of communism, as public concern focused on industrial pollution with high visibility and immediate health consequences (i.e., air, water, and soil contamination).

It was the commitment to adopt EU legislation as an accession requirement that placed chemical safety reform on the policy agenda in Central and Eastern Europe. However, what moved reforms forward and assured the smooth adoption of EU chemical standards was the interest of chemical industries in the EU and in Central and Eastern Europe in achieving harmonization of EU environmental norms. The close linkage between chemical safety rules and access to EU markets provided strong motivation for export-oriented chemical enterprises in Central and Eastern Europe to support the introduction of EU legislation.

The chapters in this part of the book follow the changes in the domestic politics of chemical regulation in the Czech Republic, Bulgaria, and Poland. Chapter 1 discusses the commercial and normative transnational pressures that reshaped the environmental interests of the chemical industry. Chapter 2 examines the evolution of chemical safety policies and the role of international commitments and industrial interests.

1

Regulatory Pressure and the Chemical Industry

In Central and Eastern Europe, chemical production is a concentrated, export-oriented, and increasingly multinational industry. As post-communist economies open to international competition and prepare to join EU institutions, the chemical sectors of these states enter highly cartelized and regulated regional and global markets. The strong trade orientation of Central and East European states to EU markets exposes chemical exporters to pressure from subcontractors, industrial associations and consumers to comply with EU chemical legislation. In addition to formal EU and OECD regulations, the international chemical industry has embraced and seeks to promote in emerging markets a range of voluntary environmental codes such as the Responsible Care Program, an initiative administered by International Council of Chemical Associations for improved environmental, health, and safety performance of chemical enterprises.[1]

The multiple environmental pressures and incentives associated with regional market and political integration have shaped the environmental interests of the chemical industry in EU accession countries and its role in the adoption of EU chemical regulations. This chapter illuminates these dynamic changes in the domestic politics of chemical safety reform by examining the influence of international markets, norms, and organizations on the policy preferences and strategies of the chemical industries in the Czech Republic, Poland, and Bulgaria.

The Czech Republic

The production of chemicals is one of the oldest and most developed industries in the Czech Republic, accounting for approximately 13 percent of industrial output and 8 percent of industrial employment in the country.[2] Czech chemical production covers all subsectors including basic chemicals, agrochemicals, paints, pharmaceuticals, soaps, detergents, fibers, and others. There are 250 chemical enterprises, but the sector is dominated by large enterprises with more than 1,000 employees, which account for more than 60 percent of the revenue in the sector.[3] After a temporary decline in the early period of the transition to a market economy, the Czech chemical industry revived rapidly after 1993, reaching 124 percent of its 1991 production by 1996 and labor productivity growth that is among the highest in the country.[4]

The Czech chemical sector relies heavily on international trade and accounts for approximately 13 percent of industrial exports of the Czech Republic.[5] Since 1990, the export performance of the sector has been strong, with the volume of exports increasing steadily. The ratio between the industry's sales for exports to total sales was 47.9 percent in 1996 and increased to 59.3 percent in 2001, with the pharmaceuticals, cosmetics, and rubber and plastics subsectors reaching ratios of 76.5 percent, 89.1 percent, and 72.4 percent respectively in 2001. The most important destination of Czech chemical exports is the EU, which in 2001 accounted for 65.8 percent of exports in the sector.[6] The chemical sector is experiencing a negative trade balance despite its strong export performance, however. A number of factors contribute to the trade deficit, including decreased demand in EU countries as a result of economic slowdown and growing domestic demand for some product groups.[7]

The privatization and restructuring of the Czech chemical industry, which began in 1992, has proceeded with delays. The majority percent assets of the industry (91.5 percent) were transformed into joint stock companies in the first stages of the privatization process. However, the state maintained significant shares in these joint-stock companies through the National Property Fund.[8] At the same time, the chemical industry is among the most attractive in the Czech economy for foreign direct investment. In the period 1990–1996, investment in chemical firms accounted for 8.5 percent of FDI in the Czech Republic.[9] International companies that have invested in chemical production in the Czech Republic include the IVAX Corporation, Pliva, Procter & Gamble, the

CONSTAB Group, Eastman, BorsodChem, and Cabot.[10] Strengthening the export position of the sector and attracting foreign investment to facilitate its modernization are considered critical for its future development and competitiveness.[11]

The Association of the Chemical Industry of the Czech Republic (ACICR) is the main societal body representing the political interests of chemical enterprises.[12] The strong international orientation of the chemical industry, its high dependence on environmentally sensitive EU markets since 1990, and the process of EU accession have profoundly affected the position of the sector and its role in environmental politics. Since its establishment in 1990, the ACICR has identified the goals of improved environmental management and European integration as central to its activities, emphasizing that a good environmental reputation and performance will improve the overall competitiveness of the sector and its access to international markets and financing.[13]

The proactive environmental position of the chemical industry association closely reflects the preferences of large chemical companies engaged in exports to the EU, which are the association's most influential members. As one ACICR executive pointed out, such companies must comply with EU regulations and voluntary standards because of "their own interest in the successful conduct of business" and their preferences and leadership determine the environmental activities and position of the sector.[14] The ACICR does recognize, however, that the application of new chemical regulations will be more costly for small and medium-size enterprises and for companies with limited contacts with EU markets. It has proposed and implemented a number of compensatory measures, such as training and implementation activities that would facilitate the application of new standards in smaller firms. Thus, the collective environmental preference of the chemical industry, although conditioned by the perceived economic costs and benefits associated with international integration and regulations, is not a straightforward function of the economic interests of individual firms. It reflects complex intra-industry bargaining, the central mediating role of industry associations, and the dominant role of large export-oriented companies, whose leverage is further strengthened by European integration.

An analysis undertaken by the ACICR with support from CEFIC and the PHARE program highlights the positive balance between benefits and costs for the Czech chemical industry as a whole associated with the adoption of EU chemical safety standards. The study estimates that the cost of implementation of EU chemical safety legislation that affects the functioning of the EU internal market would amount to 1–1.5 percent of the value of the total output of the Czech chemical industry. It also points out that additional resources will be needed for implementing regulations regarding the control of major accidents and integrated pollution prevention and control. The study emphasizes, however, the trade benefits of regulatory reform and EU harmonization: improved access to EU markets, the elimination of objections about an unfair comparative advantage and eco-dumping, considerable reduction in transaction costs associated with compliance with multiple national standards, a stable regulatory environment that would facilitate foreign investment, and avoiding "excessive" domestic regulation. The analysis concludes that the benefits of full integration in the EU market and the harmonization of chemical safety standards outweigh the cost of adapting to additional regulations, summing up the basis of the industry's preference for regulatory reform in compliance EU directives.[15] The association's comments on the 1997 draft Law on Chemical Substances before its adoption reiterate this support, arguing that "the Bill should reflect, to a considerable extent, the EU legislation in relation to the marketing of chemical substances and preparations."[16]

In addition to shaping the fundamental preferences of the Czech chemical industry for regulations compatible with the EU standards, the process of regional integration affects the strategic means through which the sector seeks to influence environmental politics. Since the early 1990s the industry has adopted an array of environmental strategies to improve its political leverage, including improved information sharing, adoption of international voluntary standards, participation in transnational business coalitions, and forceful activism in environmental policy making.

One of the first elements of the post-communist environmental strategy of the chemical industry was to build a "greener" public image. Being a symbol of unabated industrial pollution under communism, the chemical sector faced strong international and domestic incentives associated with greater demands for access to information to devote resources for improved environmental performance and reputation.[17] As part of this strategy, the ACICR collects and publishes information on the environmental investment programs of selected enterprises and on the environmental performance of the sector as a whole. This responds to public demand for improved access to information while safeguarding sensitive firm-level data. Reports of the industry association publicize the fact that the chemical sector has reduced its emissions into the air and water as well as its generation of waste more than Czech industry as whole, despite the rising level of chemical production since 1993 (figure 1.1).[18] Similar information is also published in English as a way to brighten the tarnished environmental image of the industry abroad.[19]

Although environmental data published by the industry is generally accepted with a healthy dose of skepticism, the effort of the chemical industry to collect and publicize such data is already indicative of an important change in its environmental awareness and willingness to engage in a public dialogue. In Bulgaria, for example, sector-level data on the performance of the chemical industry is still not publicly available. This reflects the relatively late start of the Bulgarian chemical

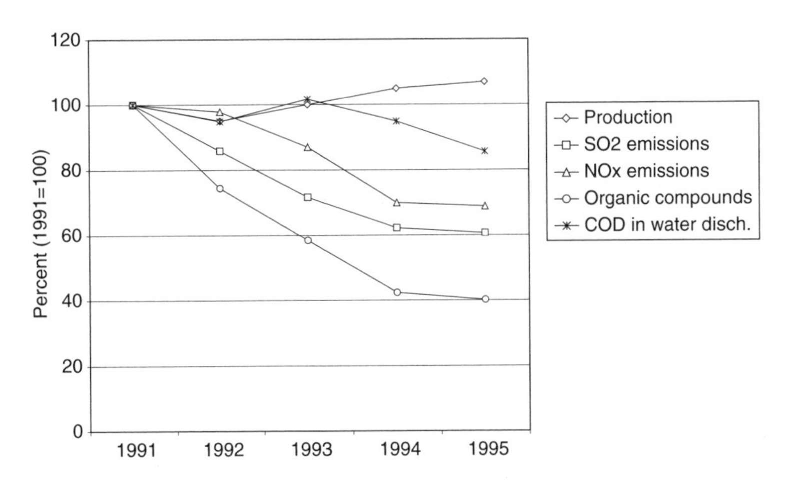

Figure 1.1
Production and pollution in the Czech chemical industry, 1990–1995. NOx: nitrogen oxides. COD: chemical oxygen demand. Source: ACICR 1997a.

industry in reforming and taking into account environmental concerns, as well as the limited resources of its chemical industry association. The lack of such data in Bulgaria considerably limits the ability of outside observers to assess the overall change in the environmental performance of the sector and the trajectory of its pollution-control effort.

Another element of the proactive environmental strategy of the chemical industry in the Czech Republic is the adoption of international voluntary standards for environmental management in close cooperation with transnational business associations. Guided by their interest in promoting international environmental norms, chemical associations in Europe, headed by CEFIC, extend considerable support for improved environmental management and the adoption voluntary norms in transition countries. The Responsible Care program in the Czech Republic, for example, was established thanks to assistance by CEFIC and advice from chemical associations from Sweden, Finland, Poland, Switzerland, Austria, and Germany. The program is now central to the activities of the chemical association, and the Czech Republic has a leading role in Central and Eastern Europe in applying Responsible Care principles. More than 50 enterprises, representing 95 percent of chemical output in the chemical industry, were involved in the implementation of the Responsible Care program by 1997. By 2001, 23 Czech companies had received the Responsible Care certificate and logo. Not surprisingly, large chemical exporters such as Chemicke Zavody Sokolov, Chemopetrol, DEZA, Kaucuk, Precheza, Spolana, and Synthesia took the lead in the establishment and application of the program.[20]

Trans-European business cooperation is an important element of the environmental strategy of the Czech chemical sector, not only in the implementation of voluntary standards but more broadly in domestic regulatory politics. Being a representative of an industry with "global interests," the Czech chemical association pursues actively political and technical collaborations with European chemical associations. It is a member of CEFIC, of the European Association of Employers in the Chemical Industry, and of the Federation Europeenne du Commerce Chimique, and cooperates on a bilateral basis with a number of West European chemical associations. Multiple documents of the ACICR emphasize the importance of such transnational contacts in the context of growing regional integration:

> The close cooperation with European non-governmental multinational organizations and also with national ones, which associate chemical companies, makes it possible to permanently present the Czech chemical industry in the world and especially in Europe. Such contacts will be made use of by the Association to support the incessant increase of the ability of the Czech chemical industry to compete in the process of integration of the Czech economy into European structures.[21]

The prospect of EU membership for the Czech Republic further deepens the international contacts of the Czech chemical industry, providing a focal point for transnational cooperation within Europe. CEFIC and the European Commission have sponsored a range of joint projects involving industry, the government, and the Commission to facilitate the participation of the chemical industry in the adoption and implementation of EU legislation in accession countries. Such initiatives include the CEFIC/PHARE project on the implementation of the chemical safety legislation related to the functioning of the single market, as well as the ChemLeg and ChemFed programs for strengthening the capacity of Central and East European chemical associations, implementation of Responsible Care, and assisting companies to adjust to relevant EU legislation.[22] The programs are funded with the support of the European Commission and implemented by CEFIC in cooperation with chemical associations of accession and EU countries. Transnational business contacts and partnerships are a source of valuable political resources for the Czech chemical association, enhancing its position in the domestic harmonization and implementation of EU legislation.[23] The advice of Western chemical associations and chemical industry experts often provide a direct input for the formulation of the ACICR's opinions on legislative drafts and its contributions to Parliamentary discussions.[24]

A similar trend of transnational business cooperation and adoption of voluntary norms is also supported by other export-oriented sectors in the Czech Republic. With the assistance of international programs and networks, a number of green business organizations were established during the 1990s, among them the Czech Environmental Management

Center, the Czech Business Council for Sustainable Development, the Czech Cleaner Production Center, and the Business Leaders Forum of the Czech Republic. In the course of the 1990s, green business associations have become influential actors in Czech environmental politics as lobbyists both in the executive branch and at the parliamentary level. They work to reshape the environmental policy space in such a way as to provide more leverage and flexibility for business in environmental politics and the adoption of EU legislation.[25] As the director of the Czech Business Council for Sustainable Development highlighted in an interview, European integration and the process of regulatory harmonization motivate a great deal of environmental awareness and activism among export-oriented companies:

> EU markets are vital for export-oriented companies and for the Czech economy as a whole, and there is significant pressure by consumers and suppliers in those markets for improved environmental performance. It is in the interest of the country to accept EU legislation in such a way as to be acceptable for industry and to establish a stable legislative environment. That is why we insist that new laws should be, as much as possible, compatible with new EU legislation. The goal is to achieve harmonization in such a way as not to damage the competitiveness of Czech industry.[26]

The ACICR is a central participant in domestic green business alliances in the Czech Republic and exemplifies the proactive position of internationally oriented industries and their effort to influence the terms of the domestic implementation of EU environmental regulations. The ACICR's Secretary for the Environment summarized the strategy of the sector in an interview: "Membership in the EU is a high priority for the Czech government, and industries would have to comply with that legislation, so it is in their best interests to consider as early as possible the adjustments that would be needed."[27] As a result of persistent lobbying and the support of domestic and international industry organizations, chemical industry representatives were appointed to inter-ministerial committees for the preparation of the Act on Chemical Substances and the Emergency Response Act. The ACICR also uses its position to influence the content of a wider range of environmental regulations—such as the Waste Management Act, the policy on Integrated Pollution Prevention Control, and air pollution provisions—which are not exclusively targeted at the chemical sector but have important cost implications for chemical production. In its policy statements, the chemical industry always emphasizes the relationship between the proposed legislation and EU law and its implication for the trade competitiveness of the sector. Where possible, the association uses the criteria of EU legislation to argue against provisions of domestic legislation entailing "excessive" regulation.[28]

In sum, the process of EU integration has provided both constraints and opportunities for strategic bargaining for the chemical industry of the Czech Republic, affecting in multiple ways its position in domestic environmental politics. The economic and environmental incentives associated with international markets motivate the sector to support stricter chemical safety regulations. The pressure and assistance from trans-European business networks provide further strategic incentives for the industry to embrace a proactive position in environmental politics. Similar trends, although played differently in the specific societal and institutional context of each country, are observable in the cases of Poland and Bulgaria.

Poland

Similarly to the case of the Czech Republic, the chemical industry in Poland is an important industrial sector characterized by a high level of export orientation and an increasing share of foreign investment. The Polish chemical sector is the largest in Europe. There were approximately 16,000 chemical enterprises in 1998, producing a range of products, including organic chemicals, fertilizers, plastics and plastic goods, household chemicals, paints and varnishes, agrochemicals, pharmaceuticals, and rubber.[29] Chemical production accounts for approximately 10 percent of industrial sales, 8 percent of industrial employment, and 10 percent of the country's industrial exports.[30] The sector experienced a temporary decline at the beginning of Poland's post-communist transition, followed by relatively rapid recovery. Since 1993, the growth in chemical sales, production and profitability has been higher than the average for Polish industry.[31]

International trade is considered a fundamental aspect of the development of the sector. Chemicals are among the ten leading industries in total trade volume for Poland. Close to 30 percent of Polish chemical production is for export, and imports are an important source of raw materials in the industry.[32] A handful of firms dominate the exports of chemical products. Twenty chemical enterprises, which are among the 100 largest Polish exporters, account for approximately 62 percent of the overall exports in the sector.[33] The EU provides the primary market for Polish chemicals. In 1995, for example, 61 percent of Polish chemical exports were directed to developed countries (mainly the EU), and 79 percent of the imports in the industry came from industrialized states.[34] In the second half of the 1990s, there was a revival of trade with the states of the former Soviet Union, making Russia one of the important trading partners for Poland's chemical industry, along with Germany, Holland, the United Kingdom, France, Italy, Belgium, the United States, and the Czech Republic. As in the case of the Czech Republic, the chemical industry in Poland experiences a trade deficit as a result of the increasing level of domestic consumption of chemical products and reliance on imported raw materials.[35]

Most chemical plants in Poland were included in the process of post-communist privatization either through commercialization or liquidation. Although the majority of chemical enterprises are private entities, a substantial portion of the chemical production is still publicly owned, as some of the largest firms have a considerable share of state ownership. According to estimates by the European Bank of Reconstruction and Development reported in a European Commission study, in 1998 about 96.5 percent of the enterprises in the sector were privately held, but the private entities' share of sector revenue was less than 30 percent.[36] Foreign investment plays an important role in the process of ownership transformation and the modernization of the sector. By 1996, foreign capital was engaged in 321 chemical companies. The most substantial foreign investments are concentrated in the production of rubber, detergents, cosmetics and toiletries, and industrial gases. Among the main investors in Polish chemical plants are big multinationals such as Unilever, Procter & Gamble, Solco Basel, Michelin, Henkel, the Boc Group, Pam-Gas, Oriflame, Beiersdorf, L'Oreal, the Cussons Group, and the Kalon Group.[37]

The internationalization of the Polish chemical industry and its strong orientation toward EU markets profoundly affected the environmental interests of the sector. The environmental performance of the sector became viewed as an integral part of its international success and an essential prerequisite for strengthening its trade performance.[38] Very early in the transition process, the Polish Chamber of the Chemical Industry (PCCI),[39] the main organization representing the political interests of the sector, included the adoption of European environmental standards among other priorities for the sector such as privatization and restructuring, improved industrial productivity, and overcoming recession.[40] The process of Poland's integration in EU markets and institutions has further shaped the environmental preferences of the sector and its choice of environmental strategies, which as in the Czech Republic include improving the industry's environmental image, participating in transnational coalitions, and adopting a proactive position in environmental politics.

In Poland, however, to a greater degree than in Bulgaria and the Czech Republic, strong domestic pressure together with international markets pushed the chemical sector to improve its environmental performance and establish a greener image. Local concern about industrial pollution and regional contamination was very high in Polish society during the late 1980s and the early 1990s, and chemical enterprises were targeted as some of the worst offenders. In 1989, the newly elected democratic government, like other post-communist governments in the region, made a commitment to place environmental issues among its policy priorities. One of the first reform measures the government undertook was the publication in 1989 of a list of the "eighty most polluting companies," which created a very visible target for societal monitoring and enforcement activity. Many of the enterprises on the government's "black list" were chemical firms. Regional lists of most polluting entities, published by local authorities, further increased the visibility of the sector as an industrial polluter.

The impact of the chemical industry on the environment became an issue that the sector had to deal with immediately. According to chemical industry representatives, the list of the eighty companies most arduous to the environment has been "menacing Poland" since 1989.[41] Such lists became financially arduous for important chemical firms. They damaged the image of companies at home and abroad, diminished their attractiveness for foreign investment, and undermined their selling prices in the privatization process. With the growing integration of the Polish economy in West European markets, it became increasingly important for the chemical industry to counter this negative image and possible accusations of "eco-dumping."[42]

In response to both domestic and international incentives, the Polish chemical industry has developed a range of strategies to improve its environmental image.[43] The PCCI uses considerable resources to promote improved environmental performance of chemical enterprises in accordance with national and international standards, and to popularize these efforts at home and abroad. The 1994 program of the association, for example, included the following priority activities: "preparation of report on [the] real impact of the Polish chemical industry on the environment, implementation of the Responsible Care Program, [and] . . . preparation of comparative study on environmental protection requirements in Poland and in the EC [European Community] together with the regulations precising responsibility of the companies for the pollution of environment."[44] The PCCI has a special Policy Advisory Team on the Environment and publishes periodic reports on the environmental performance of the industry and its largest enterprises, emphasizing the trends of declining levels of harmful industrial emissions and waste generation despite the increasing volume of production since 1991 (figure 1.2). Similar data are also presented at international forums and English language publications of the association (table 1.1).[45]

The adoption of voluntary international standards such as ISO 14,000 and participation in the Responsible Care Program is another element of the strategy to enhance the image of the Polish chemical industry. As in the Czech Republic, export-oriented and multinational firms are most eager to adopt such standards.[46] The PCCI plays a central role in promoting the broader applications of these codes and environmental management systems. In 1992, in cooperation with Western chemical industry associations and the Industrial Chemistry Research Institute, the

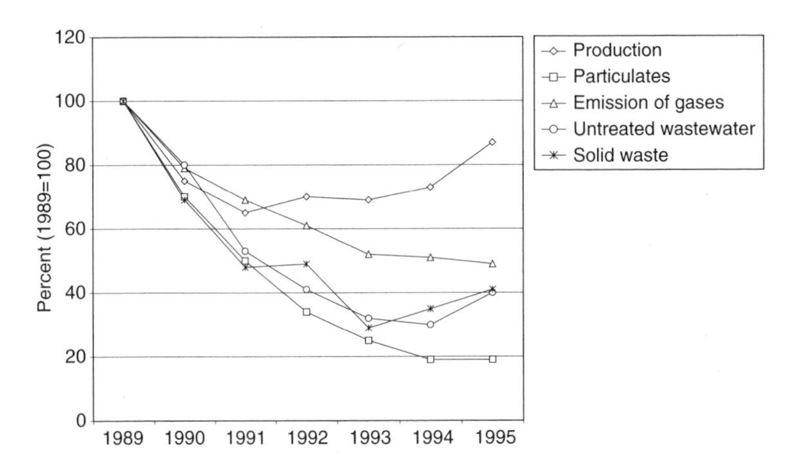

Figure 1.2
Production and pollution in Poland's chemical industry, 1989–1995. Source: PCCI 1997a.

Table 1.1
Emission reductions from the chemical industry in Poland, 1989–1996. Source: Lubiewa-Wielezynsky 1998.

Pollutant	Percent reduction
Dust	80
SO_2	25
NOx (nitrogen oxides)	25
CO	77
Wastewater volume	37
COD (chemical oxygen demand) in wastewater	62
Landfill wastes	35

PCCI adopted the Responsible Care program and undertook the development and implementation of its components. By 1997, 12 Polish companies, representing about 75 percent of the total PCCI members' sales value, participated in the program.[47] By 2001, 23 companies participated.[48] The annual reviews of the Polish Responsible Care initiative advertise the emission reductions and higher safety of production achieved by participating enterprises, the program's success in strengthening the competitiveness of chemical firms, improving the firms' relations with government authorities, and creating "a new image of chemical companies as environment friendly."[49]

The voluntarist environmental strategy of the chemical industry parallels a broader societal process of environmental image building and organization in the business community in Poland. Environmental business organizations and clean technology centers have emerged as important actors in the policy and environmental management fields. In comparison to the Czech Republic, the Polish green business network is more dispersed and often more narrowly and regionally focused. This structure is conditioned to some extent by the decentralization of the environmental regulatory and enforcement system and by the greater importance of local politics for environmental regulation. Many green business groups in Poland are local or represent a single industry rather than broad national coalitions of industrial enterprises. Some business organizations exclusively target the development of environmental management systems and clean technologies. Others operate as advocacy organizations, working both to promote the concept of sustainable development among Polish business circles and to increase the voice of industry in environmental matters.[50] There are also a number of national organizations created as branches of international business networks, including INEM Poland (an affiliate of the International Network for Environmental Management), the Association Polish Forum Club of ISO 14,000, and the Polish Business Council for Sustainable Development.[51] EU integration and participation in international markets are important factors motivating the interest in environmental management standards and the establishment of environmental business organizations.[52] As in the Czech Republic, exporters and multinational companies play a central role in such organizations, and in the adoption of clean production methods and international environmental management standards.[53]

The chemical industry is representative of this societal process of building a green image and mobilizing to respond to international and domestic environmental pressures. However, there has not been a concerted effort to build domestic cross-sector coalitions or a common front among business organizations similar to the ones established in the Czech Republic. In its environmental activities and its political strategies, the

PCCI relies on its political resources and contacts with the Ministry of the Economy and parliamentary committees to develop a lobbying strategy.

The Polish chemical industry association, like its Czech counterpart, has also established and benefited from close transnational cooperation with West European industry organizations.[54] The PCCI is a member of CEFIC and cooperates on a regular basis with subsector organizations such as Euro Chlor and the International Fertilizer Association.[55] The annual reports of the PCCI include a long list of international workshops and programs on environmental management, on-site training, trade and environment issues, and transfer of know-how.[56]

Assistance and pressure from trans-European associations, as well as the trade benefits associated with EU accession and harmonization of chemical safety standards, motivate the strong support of the Polish chemical industry for the adoption of EU chemical standards. Documents of the chemical association recognize that the adoption of EU environmental and safety standards raise concerns among industrial representatives, who fear that the additional cost of regulation may erode the comparative advantage they enjoy as a result of low labor costs.[57] However, the PCCI justifies a sector-wide preference for regulatory reform on the basis of expected net sector-wide benefits of harmonization and EU accession.

An industry analysis conducted within the framework of the CEFIC/PHARE project on the implementation of the EU chemical Acquis related to the Common Market identifies a long list of expenditures associated with the application of this legislation: generation of missing data and product information, preparation of safety data sheets and labels, updating of existing databases, testing procedures for risk assessment of chemical substances, the establishment of emergency response procedures, the cost of registration, and the procurement of employees to deal with classification, packaging, labeling, and safety data sheets. However, the study also highlights the trade benefits associated with the adoption of the EU regulations on chemical substances. These benefits include removing technical barriers to EU markets, improved market performance, reduced transaction costs, mutual acceptance of data, and improved systems of safety and control. The study concludes that "the implementation of EU requirements, after the first period of increased costs, should have a positive influence on the competitiveness of Polish chemical enterprises on the international market."[58]

A document of the PCCI also reiterates the commercial rationale behind this policy preference: "Harmonization will support international trade with chemical substances and preparations. . . . The aim should be to fully integrate the EU legislation, in order to gain maximum benefits."[59] Behind this policy position, however, lie important intra-industry differences due precisely to the differential cost implications of EU regulations for individual firms. As in the Czech Republic, EU integration and regulatory adjustment are easier and more beneficial for large companies that already operate in EU markets and implement international voluntary standards such as ISO 9,000, ISO 14,000, and the Responsible Care Program. Small and medium-size firms with no experience with international competition have less capacity to cope with new regulations and gain few if any benefits from increased economic integration. Many chemical enterprises, in fact, did not have precise information on the financial implications of regulatory harmonization with EU laws during the early 1990s.[60] However, the economic interests of big industrial exporters tend to dominate the intra-industry dynamics of policy preference formulation. The process of European integration and the government's interest in adopting the EU chemical regulations provide these interests with additional leverage in shaping the environmental-policy position of the chemical industry.

The chemical sector's proactive position in supporting the harmonization of EU chemical standards and its close collaboration with EU associations facilitate its role in strengthening the representation of the sector as a whole in environmental politics. The PCCI has also argued that the adoption of complex and costly EU standards necessitates active involvement by industry to achieve harmonization in ways that best reflect its interests.[61] The PCCI Policy Advisory Group on the Environment has used its technical capacity and its political resources to establish better working relations with relevant ministries and the parliament and to communicate effectively the industry's opinion on environmental legislation.[62] The information, expertise, and political support provided by Western chemical associations, CEFIC in particular, have further increased the voice of the Polish chemical sector. One of the main objectives of projects on the harmonization of the EU chemical legislation sponsored by CEFIC and the European Commission has been to establish an institutional framework for closer consultation among governmental ministries, EU institutions, and industry organizations. Such projects have facilitated discussions between "law makers" and "law consumers," the formulation of comments on draft legislation by the PCCI, and consideration of those comments by government authorities.[63]

The process of European integration thus affects the position and strategies of the chemical industry in Poland through multiple mechanisms. Participation in environmentally sensitive regional and global markets provides a commercial rationale for major chemical firms to change their environmental preferences and to support the adoption of EU standards that pertain to their activities. In addition, the process of EU market and institutional integration, along with targeted environmental pressure domestically, has reshaped the environmental strategies of the chemical sector, forcing it to build a positive environmental image, adopt international voluntary standards, and cooperate with transnational networks to increase the voice of industry in reforming domestic legislation in accordance with EU standards. As the next section demonstrates, similar reshaping of chemical safety politics in response to international influence is also observed in Bulgaria, although more slowly as a result of slower economic reforms and integration of the country with international institutions.

Bulgaria

The chemical industry in Bulgaria has been, since its inception, a highly concentrated and export-oriented sector. Most of the production capacity in the industry (about 70 percent of today's fixed assets) was established by the communist state in the period 1970–1985, when, on average, 70 percent of chemical production was exported, mainly to former communist countries. The sector, one of the largest in the country, comprised 108 plants and a work force of 100,000 at the beginning of the 1990s. In 1998, employment in the sector fell to approximately

43,000 people, or 6.4 percent of manufacturing employment.[64] Like other industries, chemical production experienced considerable decline at the beginning of Bulgaria's economic transition, reaching its lowest level of production in 1993 at 44.6 percent of its 1989 level of production. The recovery of the chemical sector after 1993 proceeded faster than the recovery of total industrial production in Bulgaria, although at a slower pace than in the Czech Republic and Poland, reaching 71.7 percent of its 1989 level by 1995.

The improved performance of the sector was facilitated by its strong export performance and the reorientation of chemical exports to new markets, primarily EU and Balkan countries. As a consequence of this trend, the chemical industry emerged as one of the most profitable and important sectors in the Bulgarian economy in the late 1990s. In 1998 it accounted for 10 percent of industrial production and 16 percent of Bulgaria's total exports.[65]

The process of industrial privatization and restructuring in Bulgaria began in the second half of the 1990s—late in comparison to transition countries such as the Czech Republic, Poland, and Hungary. By 1997, six chemical companies were completely privatized, and more than 50 percent of assets in the sector were privately owned by 41 chemical firms.[66] Shortly after the privatization process started, the first big foreign investment in the sector was made by the Belgian firm Solvay, which bought a majority stake in the chemical enterprise Sodi Devnya. The chemical sector is considered to be among the most attractive for foreign investment, and as such it is expected to remain a leader in the development of Bulgaria industry. Closer integration with the EU and the political stabilization of the country is therefore of great importance for the future of the sector, as such integration would enhance the industry's prospects for increased foreign investment and improved access to international markets.[67]

The environmental pressures associated with EU integration and transnational business cooperation also influence the environmental interests and strategies of the Bulgarian chemical industry, albeit in a more limited way and at a later date compared to the Czech and Polish chemical industries. This is a reflection of the relatively late start of Bulgaria's economic transition and organization of societal forces. The Bulgarian Chamber of Chemical Industry (BCCI), the main societal organization representing the sector politically, was established in 1994. It is less endowed with financial resources and has less capacity for elaborate strategy development than its Czech and Polish counterparts. The association does not publish annual environmental reports or data on the environmental performance of the sector, and it has not formulated and advertised a long-term environmental strategy for the sector that public advocates could monitor.

The involvement of industrial firms and non-governmental business organizations in environmental management and politics is a relatively new phenomenon in Bulgarian society. A study conducted by the Bulgarian Academy of Sciences in 1994 found little concern on the part of enterprises with improving their environmental image and taking a pro-environment stance.[68] The uncertainty of the economic situation, the slow pace of reforms, and growing financial deficits of the firms during the early 1990s restricted any pro-environmental initiatives. Industrial interests, represented mainly through the Ministry of Industry, were often in conflict with the goals advanced by the Ministry of the Environment. Independent societal organizations representing industrial enterprises in environmental matters were largely lacking in the early period of Bulgaria's transition.[69]

The environmental pressures of EU markets on exporting firms, growing regional integration, and cooperation with transnational business networks gradually provided important stimuli for a new environmental activism in the Bulgarian business sector. But Bulgaria did not experience such a rapid proliferation of green business organizations as Poland and the Czech Republic in the early 1990s. Through most of the 1990s, only one organization—the Clean Industry Center—represented the collective interests of industrial enterprises in environmental matters and worked to promote the adoption of clean technologies and international environmental management standards. The Clean Industry Center was established in 1995 under the Bulgarian Industrial Association, the main employer association in Bulgaria. Despite the differences in organizational and financial capacity, the center follows strategies similar to its Czech and Polish green business counterparts, seeking to raise the environmental image of Bulgarian industry abroad, to facilitate the implementation of voluntary environmental management standards, and to create an effective lobby for industrial interests in domestic environmental politics.[70] The interest and support for the Clean Industry Center initiatives is most pronounced among such export-oriented sectors as machine building, metallurgy, chemicals, and textiles.[71]

The network of national, regional, and international organizations established by the Bulgarian Industrial Association has facilitated cross-sectoral and trans-national coalition building on environmental issues during the late 1990s in Bulgaria. In the face of growing market and regulatory pressures for improved environmental performance, the chemical industry in Bulgaria is one of the sectors most interested in such collaboration. The BCCI and certain branches of the chemical industry, such as pharmaceuticals, maintain active political contacts with the Clean Industry Center. Such cooperation has been essential for raising concern among chemical enterprises for improved environmental performance and preparation for the introduction of international standards such as ISO 14,000.

In the context of weaker capacity of societal organizations and limited resources of industry organizations in Bulgaria during the 1990s, transnational cooperation programs have played an even greater role there than in the Czech Republic and Poland in building a greener and proactive environmental image of the chemical industry. The process of EU enlargement motivated a growing interest on the part of the BCCI and West European chemical associations such as CEFIC in closer interaction and support.[72] The BCCI gradually developed a network of cooperation with Western chemical industry organizations and became an associate member of CEFIC. With international assistance, the Bulgarian chemical industry committed to the adoption and implementation of the Responsible Care program in 2002, supported by leading exporters such as Solvay Sodi, Totema, Plastimo, Plasmasovi Izdelia, Agrohim, and Sofiaplast. Throughout the second half of the 1990s, the BCCI was involved in the same internationally sponsored projects on EU approximation as its Czech and Polish counterparts, including the CEFIC/PHARE program on the Impact of the Commission's White Paper on the Chemical Industry in Central and East European countries, training workshops, and the ChemLeg and ChemFed initiatives of the European Commission and CEFIC for strengthening the capacity of industry to implement EU standards. It is precisely such transnational cooperation projects that stimulated the forceful entry of the BCCI into domestic environmental politics and the approximation of EU legislation. The assistance of CEFIC experts greatly enhanced the expertise of the BCCI and the industry's influence in the process of the approximation of EU chemical safety regulations. The CEFIC/PHARE project opened more direct channels of communication among the Bulgarian chemical industry, EU industry organizations, EU institutions, and the Bulgarian government, enhancing the political leverage of the chemical association.[73]

As in the other cases examined in this chapter, it was the activity of the industry association, the BCCI, that shaped the position of the sector with respect to EU law harmonization.[74] A survey undertaken by the BCCI in 1997 demonstrated that at that time many chemical producers had only limited knowledge of EU legislation and the implications of regulatory harmonization.[75] Only a few big chemical enterprises that were trading with the EU and with OECD states had adequate information on the environmental requirements of Western markets. This situation reflected the lesser degree of Bulgaria's integration with EU markets and the slower pace of economic reforms in comparison to other states such as Poland and the Czech Republic.

Since 1997, the Bulgarian chemical association has undertaken a range of activities to raise the awareness among chemical producers of the implications of EU accession and the adoption of EU chemical safety and environmental legislation. It works to involve chemical firms more closely in the process of EU environmental harmonization through seminars on EU law and through discussions on the cost and trade implications of harmonization, as well as by eliciting the opinion of its member companies on draft chemical legislation. Most enterprises have now appointed individuals to be responsible for EU harmonization issues, realizing that these standards can have important trade and cost implications. Chemical firms have also expressed interest in establishing more permanent channels of consultation in the regulatory process, and in training programs that would facilitate the implementation of the EU system of regulations.[76]

The BCCI's efforts to raise the level of information in the industry on international chemical standards has been greatly facilitated by the expertise gathered through a number of international initiatives, including the project of the United Nations Institute for Training and Research for the Assessment of the National Infrastructure for Management of Chemicals and the CEFIC/PHARE project on the Impact of the Commission's White Paper on the Chemical Industry in Central and East European Countries (1997–1998). In the framework of the CEFIC/PHARE project, the BCCI carried out an analysis of the costs and benefits for the chemical sector associated with the adoption of EU chemical safety legislation. It estimated that the application of these regulations would result in nontrivial expenses for chemical enterprises, amounting to an estimated US$8.9 million just for existing substances.[77] As in the Polish and Czech cases, however, the Bulgarian association concluded that the benefits of integration would outweigh the costs of adopting EU chemical safety standards. It emphasized that regulatory harmonization will facilitate access to EU markets, increase the competitiveness of Bulgaria's chemical products on Western markets, and guarantee a more stable regulatory environment that would better attract foreign investors.[78]

After a series of consultations with chemical firms, government officials, and CEFIC representatives, the Bulgarian chemical association formulated a position that advocated timely and accurate harmonization of the Bulgarian chemical safety legislation with EU directives. In its policy statements, the organization argued that the reform of the national legislation on chemical substances should be a priority for the Bulgarian government. On the recommendation of its member enterprises, it also insisted on reforms in the administration system of chemical regulations. Such reforms would alleviate the regulatory burden by decreasing the administrative obstacles, thus decreasing the resources spent on the adoption of new standards.[79]

Export-oriented chemical firms have been the most important supporters of the policy position advanced by the BCCI. The interests of big firms have prevailed in shaping the policy preferences of the sector, although some smaller firms with strong interests in environmentally sensitive international markets have taken considerable steps in that direction. These companies see harmonization as a matter of necessity associated with important trade benefits and improved performance in

EU and OECD markets.[80] Most of the small and medium-size chemical enterprises, however, do not have much incentive to adopt higher international standards. Unable to afford such measures, they have argued for an extended period of implementation and compliance.[81]

The interest of the chemical sector in ensuring the adequate harmonization of EU chemical safety legislation and the support provided by CEFIC helped to place the initiative in the hands of the industry, making it an important player in the development of new chemical legislation. The BCCI established solid contacts and an influential position within the European Integration sections of the Ministry of Industry, the Ministry of Health, and the Ministry of the Environment. It participated in the inter-ministerial Working Group for the Assessment of the National Infrastructure for Chemicals Management, and it was actively engaged in preparing new chemical safety laws. The BCCI also lobbied for specific provisions and reforms that would alleviate the regulatory burden for industry, such as establishing a single governmental institution dealing with chemical substances and allowing longer implementation periods for small and medium-size enterprises.[82]

The Bulgarian case, thus, similarly to the cases of Poland and the Czech Republic, reveals a trend of increasing support by the chemical industry for regulatory reform in accordance with EU chemical safety regulations shaped by the influence of export-oriented and multinational firms and the mediating role of domestic and transnational business associations. Within a relatively short period of time, the chemical industry has emerged as an important actor in Bulgarian environmental politics. The chemical industry's participation in domestic and international business coalitions has been the most important source of leverage for the sector and the BCCI. The prospect of EU membership and environmental law harmonization provides a specific focal point for such cooperation and motivates the industry's engagement in environmental policy.

Conclusion

The case studies of the role of the chemical industry in environmental politics in the Czech Republic, Poland, and Bulgaria reveal remarkably similar trends of preference and strategy adjustment under the influence of international markets and regulatory pressures. The chemical sector in each state, driven by the interests of large export-oriented firms, has strong commercial incentives to support regulatory reform in accordance with EU chemical standards. While manifesting support for reform of national chemical laws, the chemical industries in Bulgaria, the Czech Republic, and Poland have adopted a proactive approach to the harmonization of other environmental regulations with economic consequences for the sector. The goals of such activism are to ensure the best representation of chemical interests in the process of harmonization and to minimize the costs of compliance.

The process of integration in EU markets and institutions influenced the environmental interests and political strategies of the chemical industry by providing a focal point and institutional resources for transnational coalition building and a proactive position in environmental policy making. In each of the three states under consideration here, the chemical industry has taken measures to improve its environmental image and to establish international and domestic alliances likely to enhance its bargaining position in environmental politics.

The new environmental strategies of the chemical industries of Bulgaria, Poland, and the Czech Republic, while directed by strong international influence, also reflect the specific constraints of the domestic political context. In the Czech Republic, centralized coalition building with industry organizations is an essential element of the lobbying activities and policy involvement of the chemical association. In Poland, where the making and the enforcement of environmental policy are more decentralized, the establishment of strong cross-sectoral coalitions is of lesser importance than the ability of chemical industry organizations and individual chemical enterprises to provide an effective lobby at multiple levels of government. In Bulgaria, as a result of the slower pace of economic reform and integration into Europe, the interest of the chemical industry in issues of EU enlargement and environmental reform developed later than in Poland and the Czech Republic. The Bulgarian chemical industry association has considerably smaller financial capacity and political experience than the chemical associations in the Czech Republic and Poland, and it has not yet been able to develop a sophisticated system of environmental strategies.

Despite these differences, the process of adjustment of industrial preferences and strategies in the three countries has followed broadly similar paths. This finding is quite different from the conclusions of a study by Brickman, Jasanoff, and Illegen on the development of chemical safety policies in the United States and Western Europe during the 1970s and the 1980s. Among other elements of the regulatory process, Brickman et al. examine the role of the chemical industry and conclude as follows:

> Confronted with generally similar regulatory problems, industries in the four countries [the United States, Germany, Britain, and France] have responded with widely varied strategies. . . . Differences in strategy grow from the special features of each country's regulatory system and the recognition by chemical firms that influence is maximized by exploiting every opportunity presented by the prevailing "rules of the game."[83]

In the emerging markets of Central and Eastern Europe, international rather than domestic forces and rules provide the most important constraint on the environmental strategies of the chemical industry. The internationalization of domestic chemical-safety politics, a phenomenon observable in emerging markets beyond Central and Eastern Europe,[84] reflects both the increasing globalization of chemical production and trade and the existence of a denser layer of international voluntary and regulatory norms governing chemicals.

The case studies in this chapter demonstrate that the proactive environmental strategy of the chemical industry in Central and Eastern Europe is not an isolated phenomenon. A broader set of industrial actors, characterized by a high degree of export orientation, international competitiveness, or multinational production, embrace strategies of improved environmental management, proactive involvement in environmental politics, and qualified support for EU environmental harmonization. As a result, new types of societal organizations and alliances have emerged, changing the face of environmental politics in Central and Eastern Europe. As the next chapter illuminates, the internationalization of the domestic politics of chemical reforms have profound implications for policy outcomes and for the adoption of EU chemical regulations in transition countries.

2

Chemical Safety Policies in the Czech Republic, Poland, and Bulgaria

The regulation of chemicals was a weakly developed policy area in the countries of Central and Eastern Europe under the communist system. Communist economies were geared toward state-led growth with a strong emphasis on heavy industry. Few restrictions were imposed on the chemical sector because of its economic and strategic importance. The authority to issue regulations on the production and use of chemical substances was dispersed among governmental agencies and few enforcement mechanisms were in place. There was little compatibility between the communist standards for chemical control and the ones that evolved in advanced capitalist economies.

The differences between the chemical safety systems that Central and European states inherited from communism and that of the EU are described in a 1995 study by the Regional Environmental Center (REC). The study evaluates the extent of alignment between East European and EU environmental legislation. It identifies the chemical sector as having greatest differences between the EU Acquis and Central and East European laws. The report uses a scale of 0 percent to 100 percent to compare the level of proximity of Central and East European legislation to each field of EU environmental regulations. The level of approximation in the area of chemicals, industrial risk, and biotechnology (evaluated as a single category) is among the lowest, with an average of 27 percent for all states reviewed. The study concludes that chemicals, industrial risk, and biotechnology is the "field of environmental legislation that generally has the lowest ranking among the seven environmental regulatory areas. The 27 percent compliance as an average means nothing but the beginning of the drafting of some early attempts at regulations."[1] Bulgaria, the Czech Republic, and Poland did not depart from the general trend of relatively thin regulations on chemical substances and low proximity with EU standards. The scores for these countries in the REC report were 27 percent, 31 percent, and 33 percent respectively.

In Bulgaria, chemicals were regulated partially by the Public Health Act of 1973 and governmental decrees on the import and export of certain dangerous substances. These regulations were introduced at different periods during the communist development, and there was no uniform policy on chemical safety. The system did not correspond to international standards. It was encumbered by overlapping competencies of different ministries, as well as by a great number of often contradictory licensing procedures that hampered the import, export, and marketing of chemicals.[2]

The Czech Republic, similarly, lacked a system to regulate the entire life cycle of chemicals, from production and import, to classification, risk assessment, and trade. Certain aspects of chemical control were covered by the Public Health Act of 1966 (amended in 1991), decrees of the Ministry of Health, and governmental regulations on specific substances. These legal norms, however, were incompatible with EU regulations and inadequate for free trade conditions.[3]

In Poland, the protection of health and the environment from chemical substances was reflected in general legislation. This legislation included the Act on Poisonous Substances of 1963, as well as some broad provisions of the Environmental Protection Act of 1980 and the Code of Labor of 1974. Similarly to the situation in other Central and East European countries, the Polish law did not include all aspects of chemical control. It covered only some stages in the production and use of chemicals, and differed considerably from the system adopted in OECD states.[4]

In sum, there was little legislative tradition of regulating the production, use, and disposal of chemical substances in communist states. The fragmented system of chemical control did not correspond to the EU Acquis and other international standards. Since trade in industrial products was highly concentrated within the communist block, the development of international rules for the management of dangerous substances during the 1970s and the 1980s had little influence on chemical policies of the communist countries in Europe. Chemical pollution and industrial risk became a policy issue only after the fall of the communist regimes. With the closer integration with regional markets and political structures, Central and East European governments faced the task of adopting a new, technically sophisticated and comprehensive body of chemical regulations. This was a condition for gaining membership in the EU, as well as for achieving better access to the international market for chemicals. The internationalization of domestic interest-group politics of chemical safety regulation described in the previous chapter greatly influenced the ability of accession states to radically reform their chemical safety regulations in compliance with international standards. This chapter examines the reform of chemical safety policies in Bulgaria, the Czech Republic, and Poland, illuminating the harmonizing influence of international markets and institutions.

The Czech Republic

Reform of chemical regulations was not among the environmental priorities of the newly elected democratic government of Czechoslovakia in 1990. There were more imminent problems on the environmental agenda such as immediately health-threatening air pollution, water contamination, and nuclear safety.[5] Within a short period of time, however, the adoption of chemical safety legislation compatible with international standards was included among the immediate tasks of the Czech environmental program.[6] As a result of the government's ambition for early accession to the EU and the strong interest on the part of the Czech chemical industry to improve its position in Western markets, chemical legislation reforms were promoted relatively quickly on the agenda of the Ministry of the Environment of the Czech Republic.[7] The driving force of international considerations and industrial interests was present at all stages of the development of chemical legislation: from the formulation of the governmental proposals to the parliamentary hearings and approval of the new laws.[8]

The involvement of business associations in the policy-making process, as already described in detail in the previous chapter, was motivated by their strong interest in the adoption of international standards that would facilitate trade with the EU. The main concern of the chemical industry was to ensure close approximation of EU directives and to avoid "excessive regulations"—that is, rules that would impose additional burdens on chemical enterprises without the benefit of improved access to international markets. Since the executive branch played a dominant role in drafting and negotiating the bill that regulates chemicals, it was at this level that industry representatives concentrated their lobbying efforts.

The Ministry of the Environment and the Ministry of Health were the main governmental agencies responsible for preparing the law on chemical substances. Since the first stages of the work on the new legislation, the ACICR was in contact with officials at the Ministry of the Environment and the Ministry of Health, demanding close consultation during the drafting process. In its lobbying efforts, the association relied on the support of green business organizations, the Czech Employers Union, the Chemical Workers Trade Union and the powerful Ministry of Industry and Trade. The ACICR, together with the Czech Brand Products Association, participated in the inter-ministerial Commission on Chemical Safety, which was established to negotiate policies in the area of chemical control and formed an influential lobby in the formulation and revisions of the Act on Chemical Substances and Preparations.[9]

The substantive involvement of green business organizations in environmental politics in the Czech Republic was facilitated by domestic institutional characteristics that developed early in the republic's transition period. One of the main components of the environmental reform of the new government of Czechoslovakia in 1990 and 1991 was the establishment of channels for closer consultation with social groups. Prominent environmental ministers who had strong links with the environmental movement worked to make these changes tangible, promoting a culture of openness and regular communication between public organizations and the Ministry of the Environment. Initiatives to encourage public participation included the establishment of ecological information centers, governmental support for projects of non-governmental organizations, and the organization of regular round-table discussions among ministerial officials, industry representatives and environmental groups.[10] After 1992, the relationship between the government and environmental organizations deteriorated, largely because of the priority given to economic restructuring and the low significance accorded to environmental objectives by the cabinet of Prime Minister Vaclav Klaus.[11] The environmental ministers appointed at that time were seen as extremely weak and disinterested in public consultation. As a consequence, environmental organizations directed their lobbying efforts primarily at parliamentary representatives and political parties.[12]

The framework for societal involvement created in the early transition period was once again activated in the process of EU environmental law approximation during the second part of the 1990s. Roundtable discussions on draft laws and issues related to the EU Acquis were initiated by the Ministry of the Environment and provided new opportunities for dialogue among green groups, business representatives, and the government. Non-governmental organizations were informed about new legislative developments and allowed to participate in inter-ministerial committees, thus facilitating the approximation process by providing a forum to resolve important political differences at the early stages of policy making.[13]

During the drafting of the new law on chemicals, environmental organizations as well as industry associations were invited to provide comments. However, few green groups had developed expertise in the area of chemical management and EU law approximation.[14] Only Children of the Earth Czech Republic had a strong interest in one particular aspect of chemical safety. The main concern of the organization was to develop a Pollution Release and Transfer Register (PRTR) similar to the one of the United States to be included in the Act on Chemical Substances and Preparations. The register would require the release of public information on pollution-related activities, which enterprises might not be willing to provide on a voluntary basis. Although the structure for a PRTR system was already developed in the Czech Republic, such provision was not included in the Act on Chemical Substances and Preparations. The government reached an agreement with Friends of the Earth and other interested non-governmental organizations to include the PRTR framework in a different piece of legislation later on.[15] The omission of the PRTR framework from the 1998 chemical act reflected the dominant role of industrial interests in the formulation of the new chemical safety legislation. Industrial groups strongly opposed certain aspects of the PRTR system and sought to avoid any regulations not required by the EU.

The involvement of international institutions and business networks in the Czech policy-making process also increased the pressure and the assistance for rapid and close compliance with the EU chemical management system. Czech officials working on the governmental draft of the bill on chemicals participated in international training programs and working groups sponsored by organizations such as the EU, OECD, the United Nations Environmental Program, and the United Nations Institute for Training and Research. Within the framework of international projects, the governmental draft of the Act on Chemical Substances and Preparations was presented at seminars involving officials from the European Commission, international and domestic business organizations, and government representatives. The fact that in its draft version the act was also translated into English is indicative of the great importance of international considerations and incentives for the development of this area of Czech legislation.[16]

The Ministry of the Environment and the Ministry of Health submitted the draft Law on Chemical Substances and Preparations to the Council of Ministers in June 1997. During the government discussions of the draft law, the Ministry of Industry and Trade and the chemical industry association opposed provisions that industry found disadvantageous.[17] The opinion of EU industry experts voiced to the government facilitated the chemical industry's arguments for close approximation with EU legislation. The draft was returned by the Council of Ministers for revisions before its approval as a government proposal in January 1998. The final version of the draft Act on Chemical Substances and Preparations was fully endorsed by the ACICR. In anticipation of the new regulations and the costs associated with them, the chemical industry also insisted on early notification on the content of draft implementing provisions, as well as their translation into English and publication in industry periodicals.[18]

The draft Law on Chemicals was submitted to the Czech Parliament in January 1998. This legislation was not a politically sensitive issue for parliamentary representatives since the most important constituencies had already reached an agreement at the government level of discussions. As a result, no significant changes were made in the government's proposal during the parliamentary process. The bill on chemicals (Law 157/1998) was adopted by the Chamber of Deputies in June 1998 and then approved by the Senate, despite the political crisis that erupted at the end of 1997.[19] Technical changes to the original bill were introduced in 1999 by Law 352/1999, achieving even closer approximation of EU regulations.

The Law on Chemical Substances and Preparations entered into force in its updated version on January 1, 2000.[20] The Czech legislation on chemical substances establishes procedures for risk assessment, packaging, notification, transportation, and monitoring of chemicals, and introduces a ban on some substances. Thirteen governmental regulations and ministerial decrees have been prepared as implementation provisions required by the law.[21] The Czech legislation achieves a high degree of compliance with EU chemical regulations and OECD guidelines. An EU review of the compatibility between the Czech and the EU environmental legislation evaluated the level of approximation in the chemical safety area as close to 100 percent, pointing out that the draft Act on Chemical Substances and Preparations practically copied several EU directives.[22] The screening of the new law by the European Commission also concluded that the Act is almost fully harmonized with the EU Acquis. The only differences found were in some definitions used by the draft executive decrees, which were corrected by the time of their official enactment.[23] Under the influence of strong international incentives, the Czech legislation in the area of chemical safety underwent significant reforms. Within a short period of time, the level of approximation of the EU chemical Acquis changed from a low 31 percent as of 1995 to 100 percent by the end of the decade (table 2.1). The consistent implementation of these regulations involves the support of domestic and transnational chemical associations, being actively assisted and monitored by CEFIC in the framework of the ChemLeg and ChemFed projects spon-

Table 2.1
Harmonization of EU chemical safety legislation. Sources: for 1995 levels of approximation, REC 1996a; for 2000–2001 levels of harmonization, Committee of the Environment of the National Assembly of the Republic of Bulgaria 2000, Sir William Halcrow and Partners, Ltd. 1997, Blaha 2001, Council of Minister of the Republic of Poland 2000, Andrijewski and Lewandowska 2002.

	Bulgaria	Czech Republic	Poland
EU law approximation, 1995	21%	31%	33%
New legislation	Law on the Management of Chemical Substances, Preparations and Products (2000)	Law on Chemical Substances and Preparations (1998) Law Amending the Law on Chemical Substances and Preparations (1999)	Law on Chemical Substances and Preparations (2001)
EU law harmonization, 2000–2001	100%	100%	100%

sored by the European Commission. Similar incentives for national policy adjustment to EU chemical standards underlined the process of regulatory reform in Bulgaria and Poland.

Poland

The National Environmental Policy for Poland (1991) set the main priorities and principles of post-communist environmental reform in the country. As in the Czech Republic, the task of creating a new system for chemical control did not initially appear on the priority environmental agenda. Reforms in the early 1990s emphasized regulatory decentralization, strengthening the capacity for monitoring and enforcement, and the creation of national and regional funds for environmental financing.[24] The environmental problems of greatest concern were characterized by high domestic and international visibility: industrial pollution, harmful emissions to the atmosphere, domestic and international water contamination, and waste management.[25] Chemical safety management was considered to fall under the competencies of the Ministry of Health and Social Welfare rather as a matter of environmental protection.

The reform of the national legislation on chemicals was stimulated to a great extent by Poland's obligation as an EU accession candidate to harmonize the EU environmental law. The task of strengthening the legal basis for chemical protection was first specified in the National Health Program (1993).[26] The "Executive Program of the National Environmental Policy for the Years 1994–2000," adopted by the Council of Ministers in 1994 and approved by Parliament in 1995, also established the goal of developing a new national chemical safety system and implementing it by 2000.[27] In 1997, the Minister of Health and Social Welfare issued an ordinance on substances posing a threat to human health and life. The ordinance provided criteria for classifying dangerous chemicals, stipulated ways to label these substances, and specified requirements concerning chemical safety data sheets. Although the ordinance covered important aspects of the management of dangerous substances, the Polish chemical safety system remained incomplete and inadequate for the conditions of market development and EU integration.

The Ministry of Health and Social Welfare undertook the preparation of a framework Law on Chemical Substances in accordance to EU principles. The drafting of the law was delegated to an independent organization: the Institute of Occupational Medicine in the city of Lodz. The approval of the draft by the government involved a process of coordination among a number of ministries, including the Ministry of Environment, Natural Resources and Forestry,[28] the Ministry for Labor and Social Policy, the Ministry of the Economy, and the Ministry of Agriculture and Food Economy. There was little disagreement among ministerial bodies about the need to adopt new chemical legislation, approximating as closely as possible EU and OECD regulations. The political and financial incentives extended by institutions such as the EU, OECD, the United Nations Environment Program, and the United Nations Food and Agriculture Organization provided an additional impetus for comprehensive chemical reforms. The acceptance of Poland to the OECD and the intensified preparations for EU membership negotiations further accelerated the implementation of the national strategy for chemical reform.[29]

The first governmental draft of the Law on Chemical Substances was approved by the Council of Ministers in 1997, and submitted to the Parliament in January 1998. The draft bill reflected the commitment of Poland to harmonize its legal regulations with EU and OECD standards. It specified conditions for trade and use of chemical substances as well as for the introduction of bans and production limits. Under the draft law, the Minister of Health and Social Welfare would be entitled to issue executive orders to implement specific aspects of the EU system of chemical control. The draft law also provided for the establishment of an executive agency, the Bureau for Chemical Substances and Preparations, which would undertake the administrative activities carried out in the health, environment, and industrial sectors. In the explanatory notes, attached to the formal text, the authors confirmed that the draft was based on the provisions of EU directives in the field of chemical control.[30]

The chemical industry of Poland, represented by the PCCI, was highly interested in the development of chemical legislation compatible with international standards. However, in comparison with the Czech Republic, there were initially fewer institutionalized channels for industry involvement in government decision making. Draft bills and government regulations traditionally reflected an agreement between relevant ministries, but not necessarily consultation with non-governmental representatives. The environment department of the PCCI maintained some informal contacts with ministerial bodies, but there was still no stable and close working relationship with the government administration. The PCCI relied predominantly on the Ministry of the Economy to voice the preferences of industrial enterprises in the work of intergovernmental committees. Thus, while industrial actors supported the reform of chemical legislation in the spirit of EU standards, they lobbied for greater involvement in the process of EU law approximation.[31]

The support of Western chemical industry associations and the involvement of EU institutions and industry organizations such as CEFIC in the drafting and discussion of the new chemical legislation strengthened the position of the PCCI in the policy-making process and pushed the chemical policy agenda toward closer harmonization with EU regulations. Resources provided by CEFIC and the European Commission were specifically targeted to achieve closer approximation of the EU chemical Acquis and to facilitate greater participation of business associations. In the course of the CEFIC/PHARE project on the approximation of chemical legislation included in the European Commission White Paper, the Polish government became involved in broader and more substantive consultations with chemical enterprises and the PCCI. Industry representatives had an opportunity to review the governmental draft of the new chemical legislation. On the basis of this review and the advice of EU industry experts, the PCCI formulated specific recommendations for revisions of the draft law on Chemical Substances that would better reflect the interests of chemical firms.[32] Other EU sponsored projects and consultations with experts from the European Commission and European chemical industry provided additional inputs in the formulation of the chemical legislation of Poland.[33]

In the Lower House of the Polish parliament, the Sejm, a special subcommittee was set up to work on the text of the chemicals law. The subcommittee consisted of deputies from interested parliamentary committees, independent experts, representatives of interest groups, and government officials. In September 1998, the subcommittee working on the Law on Chemical Substances organized a discussion with officials from the European Commission on the provisions of the draft legislation. On the basis of the conclusions reached during the talks with EU representatives and the objections formulated by the chemical industry, the Sejm returned the draft bill to the government to incorporate revisions that would ensure closer compliance with EU standards.[34] Although the industry objections and EU recommendations delayed the adoption of the new chemicals law, they ultimately led to a high-level approximation of the EU Acquis in the area of chemical substances.

The revised government proposal of the Act on Chemical Substances was considered fully compatible with the EU Acquis in the area of chemicals.[35] Since the goal of approximation of the EU chemical safety Acquis commanded high support among parliamentary deputies and representatives of interest groups, the revised bill was adopted in January 2001.[36] Subsequently, minor amendments were made to the bill to achieve even closer alignment with EU directives. The new law entered into force in February 2002, except for articles 8–11 concerning the functions of the Bureau for Chemical Substances and Preparations, which became effective after November 1, 2002.[37] On the basis of the new chemical legislation, the Ministry of Health prepared and adopted a range of implementing regulations concerning the statue of the Bureau for Chemical Substances and Preparations and the management of chemical substances.[38]

The ChemLeg project supported by the European Commission and coordinated by CEFIC provides further support for the implementation of the chemical law, including the training of 70 to 100 experts from enterprises in the Polish Chamber of Chemical Industry to facilitate compliance with the new legislation. The ChemFed project (also sponsored by the European Commission and supervised by CEFIC) in turn seeks to strengthen the incentives and capacity of the Polish chemical association to see to the implementation of new regulations. Technical support for strengthening the government's administrative capacity for risk assessment and the administration of chemical control is also extended by projects of the European Commission.[39] Thus, as result of very close cooperation between the government administration, EU institutions, and domestic and European business chemical organizations, Poland achieved a high level of harmonization with EU chemical safety legislation and is on track with its implementation with the support of European supranational and business organizations.

Bulgaria

In comparison to the Czech Republic and Poland, the approximation of the EU chemical safety legislation in Bulgaria started at a later date and almost entirely under the initiative of international programs and institutions. Bulgaria's early transition period was marked by political instability and economic reforms proceeded slowly. Bulgaria's prospects for EU membership were still rather distant, and the need to concentrate resources on full-scale preparations was less pronounced than in countries such as Poland and the Czech Republic, which were among the front-runners in the EU accession process.

The Ministry of the Environment and Waters[40] and the Ministry of Health shared the responsibility for regulations concerning the management of chemicals in Bulgaria. As a result of several cuts in the government's budget, these agencies had limited human and financial resources, which were allocated to areas of greater national priority.[41] The absence of chemical safety standards compatible with those of the EU hindered Bulgarian exports and became an important issue for export-oriented chemical enterprises. However, the chemical industry lacked a strong organization with sufficient political experience to formulate these interests and to influence the governmental agenda.

The Ministry of the Environment started systematic work on chemical safety issues in December 1996 within the framework of a specialized project sponsored by the United Nations Environment Program and the United Nations Institute for Training and Research. This assistance was of crucial significance for the development of chemical policy in Bulgaria. Within the framework of the project, the "National Profile to Assess the National Infrastructure for the Management of Chemicals" was created. This policy document was the first step toward the creation of a new system for chemical control. It described the whole life cycle of chemical substances that existed in the Bulgarian market, including the processes of production, import, marketing, infrastructure, and export. The project also led to the establishment of a new institutional structure: a special inter-ministerial committee, which was to meet regularly to work on problems of chemical risk and regulations.[42]

The development of chemical safety policies in Bulgaria was further accelerated by the support of the EU PHARE program and a project focusing on the approximation of chemical legislation included in the European Commission White Paper.[43] The project was initiated by the European Commission in October 1997 and was coordinated by CEFIC. The involvement of CEFIC considerably strengthened the political capacity and leverage of the BCCI and its formal contacts with governmental institutions. As a result of the project, the BCCI was placed at the center of the development of new chemical legislation.[44]

The support of the PHARE program and CEFIC was also welcomed by the ministries that shared competence in the field of chemical safety as an opportunity to receive financial and administrative resources for the adoption of EU legislation. For the Ministry of the Environment, the Ministry of Health, and the Ministry of Industry, such international projects had an important capacity-building function, strengthening their regulatory resources.[45] From the perspective of the Ministry of Industry, which represents important sectoral interests, the EU and CEFIC sponsored project had the additional benefits of contributing to the free movement of goods between Bulgaria and the EU and enhancing cooperation with industry organizations. In his letter of support for the CEFIC/PHARE initiative, the deputy minister of industry even suggested that the project would be most useful "if it finished with . . . including of [EU] Directives in Bulgarian normative documents, or in elaboration of new normative documents in conformity with European legislation. Our delay in the start of harmonization work could be compensated to some extent in a similar way."[46]

The preparation of the principles of a new framework act on chemical substances became an important aspect of the implementation of the CEFIC/PHARE project in Bulgaria. The legislative component of the project was led by the BCCI coordinator and carried out by a working group of experts, which included representatives of the chemical industry, the Ministry of the Environment, the Ministry of Health, and the Ministry of Industry. As a result of their close cooperation with Western industry experts, and on the basis of a detailed analysis of the gaps between Bulgarian and EU legislation, the working group proposed the most important elements and principles of the new chemical legislation, seeking to achieve close harmonization with the EU Acquis.[47]

The inter-ministerial committee on the management of chemicals, established within the frameworks of international projects, continued the work on the bill after the completion of the EU and CEFIC sponsored programs. The involvement of the BCCI, which distributed earlier versions of the draft law to chemical enterprises, ensured the support of the chemical sector and of the Ministry of Industry.[48] The BCCI proposed that the new law should create a single regulatory body for chemicals. Such provision was in the best interest of the chemical industry, since it would deal with problems of inefficient implementation of the law, overlapping functions of governmental authorities, and the need to obtain multiple licenses for chemical products. However, the two leading governmental agencies, the Ministry of the Environment and the Ministry of Health, opposed the establishment of a new administrative structure to deal with chemicals, defending their regulatory turf in the area. Disputes between the Ministry of the Environment and the Ministry of Health were the main cause for delays in the completion of the governmental draft of the Law on Chemical Substances.[49] Despite these administrative obstacles, however, the draft bill was submitted to the Parliament by the Council of Ministers as early as November 1999.

The support of industry groups and legislators facilitated the timely approval of the law. In the parliamentary hearing of the draft bill, the Minister of the Environment, who introduced the bill on behalf of the government, emphasized the significance of this legislation for the process of EU integration, as well as for the Bulgarian chemical industry and its exports to EU markets. The minister emphasized the high level of proximity between the proposed legislation and the EU Acquis, citing the following comment by a chemical industry expert at the European Commission on the draft law on chemicals: "If this law were the single criterion for accession to the European Union, Bulgaria should be a member already."[50]

The proposed legislation caused few disagreements on the parliamentary floor. After the introduction of some technical changes, the Law on the Management of Chemical Substances, Preparations, and Products was adopted with almost full majority in January 2000, only 2 months after being introduced to Parliament.[51] Most of the implementing regulations were adopted in 2002, specifying procedures for risk assessment and notification of new chemical substances, limitations and restrictions of the use of certain chemical substances, and regulating the import and export of dangerous substances.[52] As a consequence of strong international and industry support, the level of approximation with EU chemical safety regulations in Bulgaria increased from 27 percent in 1995 to 100 percent in the 2000–2002 period (table 2.1). Bulgaria is also participating in the ChemLeg and ChemFed projects of CEFIC and the European Commission, which target the internalization and compliance with new chemical safety standards.

The initial delay in the reform of chemical laws in Bulgaria reflected the insufficient capacity of domestic institutions and the weak representation of industrial interests. International support available both to the government and industry associations compensated to a great extent, although not fully, for these differences in national capacity. The resources provided by international agencies were of critical importance for the chemical approximation process, and had a great degree of influence on the course of chemical reforms. The chemical approximation agenda in Bulgaria followed a similar track to that of Poland and the Czech Republic, motivated by government commitment to EU integration and strong industrial interest in harmonizing chemical safety regulations with those of EU markets.

Conclusion

Comparative studies of national responses to regional and global pressures often emphasize the role of domestic institutions in shaping divergent paths of adjustment.[53] In studying the development of chemical safety policies in Central and Eastern Europe under the influence of regional integration, I also focused on the structures that underline domestic political interactions. However, in the three cases examined, institutional differences across countries had a limited impact on the process of national adjustment to EU chemical standards. The change in domestic interests and coalitions in response international influence defined the political game as one characterized by a lack of significant opposition, and as outlined in the introduction, provided conditions for the rapid adoption of EU legislation. The adoption of the EU chemical Acquis supported the governments' goals of qualifying for EU membership. It was also in the interest of the public and environmental advocates since EU chemical legislation introduced a higher level of protection from the harmful effects of chemical substances in Central and Eastern Europe. Finally, as a result of its involvement in international markets and transnational coalitions, the chemical industry adopted a strong preference for harmonization with Western standards. Table 2.1 summarizes the policy changes observed. The approximation of chemical safety regulations proceeded in broadly similar ways in Poland, the Czech Republic, and Bulgaria, with the chemical industry playing a central role in this process.

If strong opposition existed between environmental objectives and industrial interests, powerful ministries of industry or the economy could have vetoed environmental legislation, and considerably influenced the pace and strength of reforms. The preparation of legislative drafts in the executive branch was a critical part of the chemical safety regulatory process in the Czech Republic, Poland, and Bulgaria. In all three countries, laws were drafted by a leading ministry and then subjected to interministerial negotiations and bargaining. While the executive branch and civil servants have broad political constituencies, individual ministries tend to represent certain agendas and issues (for example the environment, trade, industry, etc.), as well as a bureaucratic interest in expanding their regulatory authority.

At the level of parliamentary discussions, a range of factors can also serve as veto points hampering policy reform: a coalition party with different priorities on a particular issue, legislators representing regional and district constituencies, and strong societal organizations (for example trade-unions or environmental movements). The approximation of EU chemical safety directives provoked little controversy in the parliaments of the Czech Republic and Bulgaria. Poland's parliament returned the governmental draft of the Law on Chemical Substance for revisions after the special parliamentary committee working on chemical legislation consulted with industrial representatives and officials from the European Commission. The main reason for requesting governmental revisions in the draft act was the need to ensure even closer approximation of EU legislation. Thus, formal institutional characteristics such as the number of veto players in the decision-making system and the nature of the electoral constituency of politicians did not exert considerable influence on the course of chemical safety reforms in Bulgaria, Poland, and the Czech Republic.

The other set of institutional factors examined were those associated with the organization of societal actors and their relation to the state. The analytical framework of this book underlines the importance of changing socio-economic interests as agents of international influence in domestic politics. Environmental groups in Bulgaria, the Czech Republic, and Poland share important differences as well as similarities, rooted in the history of their development.[54] In the politics of chemical safety reform, however, the similarities of Central and East European green groups seemed to be more important than the differences among them. The modern ecological movement in Central and Eastern Europe originated as a vocal opposition to communism and its consequences for the natural environment. During the period of post-communist transition, environmental groups became predominantly concerned about "green" issues such as the protection of natural sites and biodiversity, public participation and access to information, ecological education, and providing opposition to consumerism and globalization. Issues of chemical safety standards and EU law approximation were not on the immediate policy agenda of most environmental groups.[55] As a consequence, in all three states examined here, environmental groups had relatively little interest or input in the governmental and legislative procedures leading to the adoption of new chemical safety legislation.

By contrast, green business organizations and chemical associations played an active role in the reform of chemical policies. The involvement of independent industrial organizations in policy making is a new phenomenon in transition countries. During the communist period, enterprises were state-owned and the ministries of industry defended industrial objectives. As the previous chapter described in detail, within a short period of democratic development, societal organizations representing the preferences of business in environmental politics were established. Under international influence, structural relations between industrial actors and the state developed toward closer collaboration in the area of chemical safety, speeding up the approximation process in all three states.

International resources also played an important role in strengthening governmental institutions dealing with the management of chemicals, and compensated partially for national capacity differences among Bulgaria, the Czech Republic, and Poland. The financial and administrative capability of states to manage environmental problems is a factor accorded great importance in the literature on sustainable development and international environmental cooperation.[56] The limited capacity of accession states to deal with the technical complexity of the EU system of chemical control is often identified as a significant obstacle to the approximation of EU chemical directives.[57] This point was emphasized several times in an interview with the head of the chemical regulation unit of the environmental directorate of the European Commission, who emphasized that the adoption of EU chemical standards in Central and Eastern European countries is "a very complex issue, technically and strategically. These countries will need institutions, quality people, and good organization in order to take on the EU Acquis in this area. . . . I see the answer as building a capability, a good organization with quality people."[58]

Indeed, Central and Eastern European states received considerable technical and financial support targeted at improving the management of hazardous substances. In this highly internationalized regulatory area, a range of intergovernmental and non-governmental organizations worked to promote the adoption of higher chemical standards. Together with the EU, institutions such as the OECD, the United Nations Environment Program, the Food and Agricultural Organizations, and the United Nations Institute for Training and Research have contributed resources to the management of chemicals in Poland, the Czech Republic, and Bulgaria. In the case of Bulgaria, where the capability of the government to carry out chemical safety reform was the weakest, international assistance virtually introduced the issue to the national policy agenda. Such assistance contributed to the fast approximation of EU chemical Acquis and compensated for deficiencies in national capacity.

The support of international institutions, coupled with strong economic incentives associated with the functioning of international markets, thus contributed to a trend of fast approximation of the EU system of chemical safety in Central and Eastern Europe. As globalization and regional integration proceeds, such blending of international and domestic politics will become more common. And yet, this dynamic aspect of global interdependence is not sufficiently recognized. The existing scholarship on globalization and its policy effects typically examines the ways in which different domestic institutions condition varying national responses, or emphasizes the role of economic incentives and global markets as the single harmonizing influence on national policies. The globalization literature often overlooks the role of international rules and institutional structures that underpin global markets, and the possibility of a more encompassing influence of the international context on domestic politics. The three cases considered here suggest that economic interests often work in tandem with international regulatory and financial incentives to facilitate the adoption of higher international standards. This finding has important implications for international efforts that seek to promote better environmental management in developing areas of the world.

The cases in this section also present insights about the conditions under which regional and global integration would have a strong harmonizing influence on domestic environmental policies. Chemical standards affect a highly internationalized market, governed by a complex system of international rules and institutional structures. Under such conditions, important commercial and financial incentives exist for countries with export-competitive and multinational chemical industries to adopt the international system for chemical control. Similar incentives may be present in other issue areas characterized with a substantial degree of market integration and international regulations. In many other contexts, however, regional integration and international markets do not provide such strong motivation for domestic actors to embrace higher environmental standards. In such cases, international conditionality can provide a commitment mechanism for reform-minded governments. However, the course of national adjustment to outside pressures would be determined to a much greater extent by the specificity of domestic political incentives and institutions.

II

European Regulations to Combat Air Pollution

Membership in the EU implies a strong commitment to comply with EU air pollution directives and with the protocols of the Long Range Transboundary Air Pollution (LRTAP) convention. Central and Eastern European states are both recipients and exporters of significant amounts of cross-border air pollution. Cooperation between Eastern and Western Europe to limit the transboundary flows of acidifying pollutants dates back to the 1975 Conference on Security and Cooperation in Europe, motivated largely by the interest of Soviet leaders at that time in using environmental issues as a platform to advance the process of détente. The LRTAP convention was singed in 1979 under the auspices of the United Nations Economic Commission for Europe (UNECE). This was the first East-West environmental agreement. Despite the broad provisions of the convention, it established a framework for continued monitoring, research, and the negotiations of more precise regulations.

In the 1980s and the 1990s, a number of protocols to the LRTAP convention were signed. The First Sulfur Protocol (1985) committed parties to a uniform 30 percent reduction of 1980 SO_2 emission levels. The negotiation of the Second Sulfur Protocol (1994) was based on the concept of "critical loads" and established differential national obligations guided by the effects of acid depositions on ecosystems and human health (table II.1). (A critical load is defined as "the highest load that will not cause chemical changes leading to long-term harmful effects on the most sensitive ecological ecosystems" (Levy 1995, p. 61).) The Second Sulfur Protocol also sets permission limits for combustion sources, and a requirement to use best available technology not entailing excessive costs. (See UNECE 1994 and table II.2.)

Table II.1
The 1994 Second Sulfur Protocol: commitments for national emission reductions (percent of 1980 emission levels). Source: UNECE 1994.

	2000	2005	2010
Austria	80		
Belarus	38	46	50
Belgium	70	72	74
Bulgaria	33	40	45
Canada	30		
Croatia	11	17	22
Czech Republic	50	60	
Denmark	80		
Finland	80		
France	74	77	78
Germany	83	87	
Greece	0	3	4
Hungary	45	50	60
Ireland	30		
Italy	65		
Liechtenstein	75		
Luxembourg	58		
Netherlands	77		
Norway	76		
Poland	37	47	66
Portugal	0	3	
Russia	38	40	40
Slovakia	60	65	72
Slovenia	45	60	70
Spain	35		
Sweden	80		
Switzerland	52		
Ukraine	40		
United Kingdom	50	70	80
European Community	62		

Table II.2
Emission limits for stationary combustion sources set by Second Sulfur Protocol. Source: UNECE 1994, Annex V.

	Thermal capacity (MW)	Emission limit value (mg SO_2/Nm^3)[a]	Desulfurization rate (%)
Solid fuels	50–100	2,000	
	100–500	2,000–400[b]	40 (for 100–167 MW) 40–90 (for 167–500 MW)
	>500	400	90
Liquid fuels	50–300	1,700	
	300–500	1,700–400	90
	>500	400	90
Gaseous fuels			
Gaseous fuels in general		35	
Liquefied gas		5	
Low calorific gases		800	

a. Nm^3: normal cubic meter.
b. Here and in later tables, the representation in the protocol is reproduced.

Other protocols adopted under the convention include the 1988 protocol on NOx, the 1991 protocol controlling emissions of volatile organic compounds (VOCs), the 1998 protocols on heavy metals and persistent organic pollutants, and the 1999 multi-pollutant Protocol to Abate Acidification, Eutrophication, and Ground-level Ozone. The 1999 multi-pollutant protocol addresses simultaneously several environmental problems on the basis of improved scientific understanding of the interaction between air pollutants and their multiple effects. It specifies new national reduction targets for sulfur, nitrogen oxides, VOCs, and ammonia, and it sets limit values for emission sources (UNECE 1999a,b).

The legislation of the EU on air pollution developed in parallel with the pan-European efforts to control transboundary flows of acidifying compounds. The EU Acquis in this area includes three types of standards: ambient concentrations for specific pollutants, emission limits for industrial sources, and product standards. After the negotiations of the LRTAP

Table II.3
Emission limits for new sources set by the 1988 Large Combustion Plant Directive. Source: European Council 1988, Directive 88/609/EEC, Annex III–VIII.

	Thermal capacity (MW)	SO_2 (mg/Nm^3)	Desulfurization rate (%)	NOx (mg/Nm^3)	Dust (mg/Nm^3)
Solid fuels	50–100	2,000		650	100
	100–500	2,000–400	40 (100–167 MW)	650	100
			40–90 (167–500 MW)		
	>500	400	90	650	50
Liquid fuels	50–300	1,700		450	50
	300–500	1,700–400 linear decrease		450	50
	>500	400		450	50
Gaseous fuels				350	5
Gaseous fuels in general		35			
Liquefied gas		5			
Low calorific gases		800			

convention, the European Community adopted its directive on sulfur dioxide and suspended particles (80/779/EEC), which was intended to deal with acidification and the adverse health effects of air pollution. EU legislation also specifies concentration limits and guide values for nitrogen dioxide and for lead. EU product standards that seek to limit harmful emission into the air apply to the sulfur content of certain fuels, to the concentration of lead and benzene in petrol, and to emissions from motor vehicles.

In the late 1980s, after the signing of the First Sulfur Protocol and shortly before the negotiation of the Second Sulfur Protocol, European Community members agreed on the Large Combustion Plant Directive (88/609/EEC—see European Council 1988). The Directive applies to combustion plants with thermal input of 50 megawatts or more. It specifies SO_2, NOx, and dust emission limits for new plants (table II.3), national ceilings for total emissions from exiting plants, and a requirement for member states to draw and implement programs for the progressive reduction of annual emissions. (Directive 88/609/EEC defines "new plants" as combustion plants for which the original construction or operating license was granted after July 1987. See European Council 1988.) As tables II.2 and II.3 show, the sulfur emission limits for large combustion sources are almost identical in the 1988 EU directive and the 1994 Second Sulfur Protocol of the pan-European LRTAP convention. The most important difference between these two sets of sulfur emission regulations is that, whereas the EU directive regulates only sources built after 1987, the emission limits of the Second Sulfur Protocol become effective for new as well as existing plants after 2004. The 1988 Large Combustion Plant Directive was amended in 2001, further tightening the emission standards for combustion sources.

Because of the high level of international environmental externality of air pollution emissions, EU member states have exerted strong pressure on Central and Eastern European accession candidates to reduce their contribution to transboundary flows of acidifying substances. Sweden, Denmark, Germany, Finland, and Austria, for example, receive a large share of their air pollution from Central and Eastern Europe and are highly concerned with the ability of post-communist states to fulfill their international obligations. At the same time, the Nordic countries and

Germany are important allies of Central and Eastern European states in their quest for EU membership, as they are among the strongest proponents of enlargement. Whereas during earlier accessions Greece, Portugal, and Spain were able to negotiate an exception to the Large Combustion Plant Directive and did not sign the First Sulfur Protocol (Churchill et al. 1995), from the perspective of EU members and the European Commission such derogations are not acceptable for Central and Eastern European applicants.

After Sweden, Denmark, and Austria became members of the EU, there has been a renewed activism within the European Union to reduce further acidifying emissions and their effects on health and the environment. In 1997, the European Commission presented to the European Council a proposal for a Community Strategy to Combat Acidification. The 1997 Acidification Strategy included ambitious national emission ceilings for SO_2, NOx, VOCs, and ammonia, an amendment of the Large Combustion Plant Directive to introduce new emission limits for SO_2 and NOx for new plants; and a new directive limiting the sulfur content of liquid fuels. (The new directive on the reduction of the sulfur content of certain liquid fuels, 1999/32/EC, was adopted in 1999. Directive 2001/81/EC, which set national emission ceilings for atmospheric pollutants, was adopted in 2001, and the new Large Combustion Plant Directive was adopted in the same year.) Other important components of the initiative were the ratification of the Second Sulfur Protocol and the reduction of acidifying emissions in Central and Eastern Europe as a precondition for achieving the new acidification targets within the EU. With this in mind, the EU Acidification Strategy emphasized that in the context of accession preparations and negotiations, the European Commission would pay close attention to air emission reductions as a priority area for discussion (Commission of the European Communities 1997c; *Enlarging the Environment* 1997).

This part of the book examines how Bulgaria, the Czech Republic, and Poland and their electricity industries dealt with the task of complying with the Second Sulfur Protocol and the 1988 Large Combustion Plant Directive during the first decade of post-communist transition (1990–2000). These European regulations require accession countries to reduce total emissions of SO_2 significantly and to ensure that combustion sources meet specific emission and technology standards. I focus on this subset of European air pollution regulations because they affect an important industrial sector: electricity generation, which supplies predominantly domestic consumers and which during the 1990s was not highly integrated in European markets.

In comparison with chemical safety regulations, which are the subject of part I, the reform of air emission regulations toward compliance with international standards presents a different set of economic and political incentives associated with EU integration. The reduction of acidifying emissions would enhance the welfare of Central and Eastern European societies as a whole by reducing air pollution harmful to health, and it corresponds to the policy objectives of environmentalists and the government. However, for the main regulated actor, the electricity industry, strict air emission standards impose high costs. These costs are not offset, at least in the short run, by any significant benefits from trade or EU integration. This selection of industries and regulations thus provides an opportunity to evaluate the argument of the differential effect of EU environmental conditionality across domestic industrial actors characterized with a different degree of integration and competitiveness in EU markets.

Electricity generation in Central and Eastern Europe and in most countries of the EU was highly protected and strongly oriented to domestic markets through most of the twentieth century. A process of gradual liberalization of the EU internal electricity market and greater energy cooperation was initiated in the early 1990s, but it proceeded with delays. The European Commission published its first paper on the creation of a single energy market in 1988. In 1992, it formally submitted a proposal specifying a set of rules for a common market in electricity and gas. This initiative met with considerable resistance from member states whose electricity supply was concentrated in vertically integrated utilities that enjoyed a great degree of protection. The EU Electricity Directive (96/92/EC) was adopted in 1996 after prolonged negotiations (Cross 1996; Eising 2002).

The 1996 EU Electricity Directive seeks to increase transparency and competition in the electricity sector by specifying authorization and tendering procedures for new generation capacity, by allowing negotiated or regulated third party access to national grids, and by unbundling the accounts of electricity production, transmission, and distribution. The directive also provides for a gradual liberalization of national electricity markets to achieve a market opening of about 33 percent by 2003. The provisions for the liberalization of European electricity markets, however, can be challenged on the basis of the public service obligation clause, which member states may evoke in the interest of the security, regularity, quality, and price of supplies, as well as for purposes of environmental protection (European Council 1996).

Simultaneously with these efforts to create an internal energy market, the EU also increased energy cooperation with countries from Central and Eastern Europe. These initiatives include the Synergy program for energy cooperation with third countries, participation in the European Energy Charter and its protocols, and the establishment of interconnection between the electricity grids of Central European states and the Union for the Coordination of Production and Transmission of Electricity (UCPTE), which links all West European electricity grids. As integration and liberalization of energy markets in Europe proceed, electricity companies in Central and Eastern Europe are likely to seek to increase their share of exports and access to EU markets. The restructuring of the electricity sector in post-communist economies, which intensified at the end of the 1990s partly to meet the requirements of the EU electricity directive, is also likely to increase the share of international capital in this sector.

The closer integration of European electricity markets and the gradual penetration of international capital in the energy sector of Central and East European states implies that in the future EU markets and environmental rules are likely to exert a stronger and more direct influence on the environmental interests and political strategies of the sector. Such developments would provide additional evidence of the validity of the argument that a higher level of regional and international market integration is likely to increase the sensitivity of Central and East European industries to EU regulations and their willingness to support harmonization of EU standards. (The Czech case, discussed in chapter 3, shows that such trends were already starting to take shape in the Czech electricity sector at the end of the 1990s as a result of the increasing share of foreign stockholders in the Czech Electricity Company and its growing appetite to participate in the European electricity markets that started to open at that time.)

During most of the 1990s, however, the electricity industry in Central and Eastern Europe remained predominantly oriented to domestic markets. In the three states examined in this book, exports represented 1.6–10 percent of annual electricity production in Bulgaria, about 4–9 percent of electricity production in the Czech Republic, and about 7–8 percent of electricity production in Poland (CEZ 1997, 1998, 1999; NEK 1999; Ministry of the Economy of the Republic of Poland 1997, 2002). The electricity industry of Central and Eastern European states thus gained few if any commercial benefits associated with EU integration and the adoption of EU environmental rules. On the contrary, EU and LRTAP air emission standards implied high costs for the electricity sector in accession states associated with the installation of pollution-abatement equipment during the difficult period of structural adjustment. The comparative study of air pollution reform in Bulgaria, Poland, and the Czech Republic thus enables us to identify the political incentives behind different national responses to international environmental commitments that impose high costs on domestic actors without offsetting benefits.

The three chapters in this part of the book examine the domestic political and institutional contexts that underpin the making of environmental policies in the Czech Republic, Bulgaria, and Poland. Each chapter documents the constellation of political interests, the evolution of the strategies of the electricity industry under the shadow of international commitments, and the reform and implementation of national air pollution legislation.

3

The Czech Republic: Early Adaptation

In the aftermath of communist rule, the Czech Republic, which was the more industrialized part of Czechoslovakia,[1] emerged as one of the most polluted areas in Europe. The high concentration of heavy industry and power production on the territory of the republic was causing severe environmental damage. Air pollution was among the most pressing environmental issues. The problem peaked in the 1980s, when millions of tons of sulfur dioxide and dust were emitted annually from Czech industrial sources, affecting adversely public health and the natural environment. Czechoslovakia as a whole was one of the biggest exporters of pollution to neighboring and more distant countries. Similar to other states in Central and Eastern Europe, the newly established democratic government of Czechoslovakia faced strong domestic and international pressures to address the problem of air pollution.

The policy response in Czechoslovakia and subsequently in the Czech Republic was swift. In 1991, the federal Czechoslovak government adopted a strict Act on Clean Air, following international standards. This federal legislation was adopted by the Czech Republic after the split of Czechoslovakia in 1992 and was implemented by the end of the decade. The electricity sector made significant investments in pollution-abatement technologies. As a result, major reduction in the emissions of air pollutants was achieved throughout the 1990s exceeding international standards.

Here a puzzle arises. How was the ambitious air pollution reform of the Czech Republic and its over-compliance with international standards possible given the high cost of regulation for the powerful electricity sector? This sector is traditionally a strong lobby in former communist economies where it was considered for decades to be the backbone of development. The experience of industrialized states as well as of former socialist countries demonstrates that utilities are often able to resist, delay, or weaken the force of air pollution legislation. The Czech experience is very different from that of a number of other Central and Eastern European states, among them Bulgaria and Poland, where electricity utilities have been much more successful in opposing the adoption and implementation of costly air emission standards. What accounts for the Czech strategy of early adoption of international norms?

This chapter examines the political and institutional factors that facilitated the adoption of strict clean air legislation in the Czech Republic and provided incentives for its implementation. The analysis pays special attention to the evolution of the environmental strategies of the electricity sector and the political incentives that conditioned the early adjustment of industrial strategies.

Favorable Political Context

The first democratically elected government of post-communist Czechoslovakia is often described as a government of dissidents.[2] Vaclav Havel, a dissident playwright who commanded high moral authority at home and abroad, was elected president of Czechoslovakia at the beginning of 1990. In June 1990, the two political groups that had initiated the democratic revolution—the Civic Forum and its Slovak counterpart Public Against Violence—won an overwhelming majority (57 percent) in the elections for the Federal Assembly. The Civic Forum also gained close to 50 percent of the seats in the Czech National Council, while Public Against Violence won 29 percent of the seats in the Slovak National Council, followed by the Christian Democratic Movement with 19 percent, and by the Slovak National Party and the Communist Party of Czechoslovakia, each with 14 percent. The Civic Forum–Public Against Violence coalition formed the new federal government, which included a number of popular figures associated with the communist opposition.

As the slogan of the Civic Forum proclaimed, "the return to Europe" was one of the main elements of the electoral platform of the coalition, together with the establishment of democratic pluralism and a market economy. The perceived requirements for future accession thus exerted influence on the policy choices of Czechoslovakia, and of the Czech Republic in particular, long before the European Community established any specific membership requirements.[3] The Czech Republic became one of the leaders on the road of economic and political reforms and the quest for EU accession. Its economy recovered relatively fast from the initial post-communist downturn (table 3.1), and in December 1995 it became member of the OECD along with other industrialized countries.

Environmental cleanup and cooperation were viewed by the first post-communist government of Czechoslovakia as important for improving the international image of the country and its status as a good citizen of Europe. The Federal Committee for the Environment, the Czech Ministry of the Environment, and the Slovak Commission for the Environment were created at the beginning of 1990. Shortly after their establishment, the three environmental ministries started close collaboration with the World Bank and the European Community on a joint environmental study. The study defined air protection as a target for immediate action, highlighting also solid waste and water contamination problems.[4] The prime minister of Czechoslovakia also identified environmental considerations as central to the program of his administration. The reduction of air pollution from power stations was going to be the first most important target.[5] The emphasis on air pollution reflected its high international visibility as a problem, as well as strong public concern with air pollution health impacts.

The fast development of industry and increasing output of electricity during the communist development of Czechoslovakia had caused visible and rapid deterioration of air quality. Electricity production was based primarily on low-grade brown coal with high content of sulfur. Emissions of sulfur dioxide reached a record of 3 million tons a year in the 1980s, making Czechoslovakia one of the heaviest polluters in Europe. The adverse effects of air pollution were exacerbated by the uneven concentration of industrial facilities and power production in the northern part of the Czech Republic. Northern Bohemia, together with neighboring areas in the former East Germany and in Poland, formed in the heart

Table 3.1
Czech Republic economic indicators. Source: World Bank 2002.

	1990	1991	1992	1993	1994	1995	1996	1997	1998	1999	2000
GDP growth (%)	—	−11.6	−0.5	0.1	2.2	5.9	4.3	−0.8	−1.2	−0.4	2.9
GDP per capita (1995 US$)	5,270	4,682	4,654	4,651	4,752	5,037	5,261	5,226	5,168	5,157	5,311
Inflation (%)	—	—	—	—	10	9	9	9	11	2	4
Unemployment (%)	0.7	4.1	2.6	4.3	4.3	4	3.9	4.8	6.5	8.7	8.8

of Europe the "Black Triangle," a region named after its devastated landscape and environment. The average annual sulfur dioxide concentration in a number of Czech cities situated in Northern Bohemia and the Ostrava region increased by more than 100 percent from 1970 to 1985, reaching levels that were sometimes four times the average for OECD cities.[6]

The high emissions of sulfur in the Czech Republic, together with emissions of dust and carbon monoxide, had a direct negative impact on public health in the most polluted areas, as indicated by comparative figures for cancer, circulatory problems, respiratory diseases, and infant mortality. A 1993 World Bank report on the environment and health in Eastern Europe cites a number of studies undertaken by Czech institutes that demonstrate consistently a correlation between higher concentration of air pollutants and health problems. One of these studies shows, for example, that in districts with the highest concentration of air pollution, the risk of infant mortality increased by 38 percent, the risk of post-neonatal mortality increased by 61 percent, and the risk of post-neonatal mortality caused by respiratory problems increased by a factor of seven. Another study, undertaken among school children between 1982 and 1984 in Central Bohemia, also showed that the incidence of respiratory diseases such as sinusitis, tonsillitis, bronchitis, asthma, influenza, and pneumonia was on average more than twice higher in highly polluted cites than in the control areas. In general, there has been higher incidence of respiratory disease, childhood retardation, infant mortality, and lung and stomach cancer in highly polluted areas, and especially in the coal-mining industrial districts of Northern and Central Bohemia.[7]

Acidifying emissions also caused wide spread damage of forests and cultural monuments. According to 1990 estimates by the Czech government, "affected areas," where conditions of the environment were critical, represented 10 percent of the territory of the Czech Republic and 40 percent of its population.[8] The sheer scale of devastation made the environment, and air pollution in particular, a high priority for Czech citizens immediately after the fall of communism. A Gallup poll taken at the beginning of 1990 found that more than 80 percent of the respondents considered that finding a solution to the environmental crisis was the most pressing issue to be tackled by the government, even ahead of economic reforms.[9] There was also wide public support for the environmental movement as indicated by the relatively high rating of the newly established Green Party immediately before the elections of June 1990. On the basis of opinion polls, the Green Party expected to win 8–10 percent of the national votes. The actual electoral performance of the Greens was much weaker, gaining only 4.1 percent of the Czech and 3.2 percent of the Slovak vote, and thus failing to meet the 5 percent threshold to enter Parliament.[10]

Air pollution and its health and ecological effects were also among the priorities of the Czech environmental movement in the late 1980s and the early 1990s. In the years before the democratic changes, the struggle for a healthy environment was seen as an integral part of the struggle for liberty and human rights. Dissident as well as some official environmental organizations worked to provide information about the extent of environmental damage during the communist period. After the democratic transformation, the environmental movement in the Czech Republic was in a favorable position as many activists had entered governmental or parliamentary positions, establishing permanent communication and personal contacts between the administration and societal groups.

The degree of overlap between the new political regime and dissident organizations was particularly evident in the Ministries of the Environment. Both the Federal Minister Josef Vavrousek and the Czech Minister Bedrich Moldan were former members of the Ecological Section of the Czech Biological Association, an organization that had been actively publicizing information on the state of the environment during the communist era. After the democratic changes, Vavrousek maintained strong and direct links with the environmental movement. The Federal Minister chaired the umbrella association Green Circle, created in 1989 to strengthen the political leverage and coordination among environmental groups. The Green Parliament was founded in 1991 with the support of the Czech Minister of the Environment as an institutional channel for regular communication between the Ministry and environmental groups.[11]

The political identity of the first post-communist government of Czechoslovakia, its policy objectives for comprehensive reforms and rapid integration with Europe, and the strong public support for environmental improvement created favorable political conditions to address air pollution—a problem of great domestic significance and high international visibility.[12] The Rainbow Program, the first national environmental policy of the Czech Republic, identified the reform of air pollution policies as the highest priority because of its impact on the health of the population and because of its international significance: "An important task will be the fulfillment of Czechoslovak international obligations issuing from the agreement on long-distance atmospheric pollution.... Until now measures to reduce SO_2 and NOx emissions have been inadequate and ineffective, and the non-fulfillment of our obligations casts doubt on the possibility of our proclaimed return to Europe."[13]

The Rainbow Program also specified the regulatory mechanisms to deal with local and transboundary air pollution: "strong legislation for the atmosphere and the consistent application of this legislation [based] on experiences in advance industrial countries."[14] The regulatory approach projected by the Czech Ministry of the Environment thus emphasized strict command-and-control regulations intended to induce environmental investments and a change in the behavior of industry.[15] The Czechoslovak Ministries of the Environment acted on this task quickly. The Act on Clean Air, which is one of the most comprehensive air pollution laws in Europe, was passed within only a year after the appointment of the new democratic government.

The Act on Clean Air

The Act on Clean Air (1991) was the first important piece of environmental legislation adopted by the government and approved by the Czechoslovak Federal Parliament after the democratic changes.[16] The new air pollution legislation was adopted even before the framework Act on the Environment,[17] which is indicative of its political importance. The work on the clean air legislation started early in the transition period under the initiative of the Czech Ministry of the Environment. The team included experts from the Slovak Commission and from the Federal Committee of the Environment to ensure the formulation of a draft that addressed the concerns of both republics as well as of the federal government.

The command-and-control approach adopted in the law and the stringency of the emission standards reflected the preferences and priorities of the environmental ministries, which were closely connected to and influenced by the environmental movement. The new law was designed after the air pollution legislation of Germany, which is one of the strictest in Europe. The ministerial group working on the air pollution act also made a concerted effort to align its provisions with the legislation of the European Community.[18] The Czechoslovak government approved the draft of the Act on Clean Air without any significant disagreement among concerned ministries. Very importantly, the Ministry of the Economy, created after the democratic changes to combine the functions of the former Ministry of Metallurgy, Ministry of Fuels and Energy, and Ministry of Agriculture and Food, offered a high degree of support for the new clean air legislation. This inter-ministerial agreement reflected the strong commitment of the new government to tackle the most urgent environmental problems in ways compatible with West European standards.[19]

Once approved by the federal government, the Act on Clean Air was introduced in the Federal Assembly. The parliamentary majority, which had a decisive influence on the makeup of committees, shared the environmental objectives of the executive branch. Prominent environmental activists, including Federal Minister Vavrousek, also lobbied parliamentary committees and members of parliament to ensure the timely approval of the law. During the sixteenth session of the Federal Assembly, Vavrousek presented the governmental proposal for discussion and voting. In his speech, he emphasized the devastating effects of air pollution on public health and the economy, as well as the importance of achieving international standards as a step toward Czechoslovakia's return to Europe.[20] These themes resonated with the political preferences and ideological beliefs of the parliamentary majority, whose environmental objectives were often more radical than those of the government.

Parliamentary representatives from Northern Bohemia and Moravia provided strong support for environmental cleanup, despite the fact that local industries were going to bear a significant share of the cost of complying with tough new air pollution standards. Big power plants were considered the chief culprits of the dire environmental conditions in these regions.[21] The strict air emission standards proposed in the new Act on Clean Air were penalizing the electricity generation utilities. At the same time, the Act on Clean Air intended to induce substantial new investments for coal cleaning technologies and to preserve a strong reliance on local coal for electricity production, while minimizing its environmental impact. Thus, in 1991, political forces in the Czechoslovak parliament were largely in favor of strict air emission standards, reflecting the activist position of many parliamentary members and the high concern with environmental quality among their constituents.[22]

The Act on Clean Air was passed on July 9, 1991 to replace the 1967 Act on Air Pollution. The new legislation entered in force on October 1, 1991. Measures (1) and (2) of the act, issued on October 1, 1991 categorize pollutants and set emission limits in accordance with European Community standards and guidelines by the World Health Organization (table 3.2).[23] The Act on Clean Air defines pollutants that will be regulated, establishes ambient air quality standards, and specifies emission limits for industrial entities as well as for mobile sources. Stationary combustion units are categorized into small (with output less than 0.2 MW), medium (0.2–5 MW), and large (over 5 MW). This classification is even more demanding than EU regulations and the protocols of the LRTAP convention, which define large sources as combustion plants with capacities of 50 MW or more.

The strict emission targets specified by the Act on Clean Air were effective immediately for new combustion sources, and were to be applied to existing sources within 5 years, a deadline that was subsequently extended to December 31, 1998. In the spirit of the 1988 Large Combustion Plant Directive of the European Union, existing sources were also required to draw programs for the reduction of emissions, and to coordinate with state authorities compliance timetables. Similar to the Large Combustion Plant Directive and the Second Sulfur Protocol, the Czechoslovak Act on Clean Air mandated the use of best available

Table 3.2
Air emission standards for large combustion sources in the Czech Republic. Source: Act on Clean Air of the Czech and Slovak Federal Republic (309/1991), Appendix 3.

	Thermal capacity (MW)	Solid pollutants (mg/m^3)	SO_2 (mg/m^3)	Nox (mg/m^3)
Solid fuel	5–50	150	2,500	650
	50–300	100	1,700	650
	>300		500	650
Smelting boilers				1,100
Liquid fuels	5–50	100	1,700	450
	50–300	50	1,700	450
	>300	50	500	450
Gaseous fuels	5–50	10	35	200
	50–300	10	35	200
	>300	10	35	200

technology not entailing excessive costs. The law also authorized special restrictions for periods of severe smog conditions. Thus the Act on Clean Air of 1991 provided a comprehensive basis for the regulation of air pollution issues and included the specification of ambient, emission, and technology requirements. This approach was different from the strategy of Bulgaria and Poland, the other two countries examined in this book, which first adopted a general legislation setting the principles of air protection and then issued executive ordinances to set specific standards.

After the split of Czechoslovakia, the Czech Republic adopted the Act on Clean Air along with other Czechoslovak federal legislation. Executive decrees of the Czech Ministry of the Environment regulate a range of more specific issues such as record keeping and provision of information from large and medium sources, regions requiring special air protection, the quality of fuel for combustion, and methods for measuring emissions from pollution sources. The regulatory system of air protection in the Czech Republic also includes the 1991 Act of the Czech National Council on the State Administration of Air Protection and Charges for Pollution. The act defines the competencies of administrative bodies in the Czech Republic, and specifies pollution charges and

Table 3.3
Expenditures of the State Environmental Fund of the Czech Republic (1995–2000, million US$). Data in million CZK provided by REC 2001, converted into US$ by author using average annual exchange rates provided by the Czech National Bank.

	1995	1996	1997	1998	1999	2000
Total	183.8	169.6	103.9	69.1	73.6	66.0
Air pollution	89.6	84.0	38.0	28.1	30.7	30.9
% air pollution	49	50	37	41	42	47

enforcement mechanisms. It establishes the Czech Environmental Inspection, which together with its regional offices is responsible for monitoring and enforcing air pollution standards for all major sources.[24]

Considerable resources were put into strengthening the system of fees and fines as well as monitoring of air pollution in the Czech Republic.[25] Automatic monitoring stations now provide information on ambient concentrations of air pollutants. The national emission balance is based on the Register of Emissions and Air Pollution Sources (REZZO), which includes four types of emission inventories: REZZO 1 which covers large stationary sources, REZZO 2 for medium stationary sources, REZZO 3 for small sources, and REZZO 4 for mobile sources. The amount of emissions from stationary sources is also divided and monitored by region, providing an up-to-date picture of the relative level of pollution affecting different administrative areas of the Czech Republic.[26]

The State Environmental Fund of the Czech Republic was established in 1991. The income of the fund consists mainly of revenues from fines and penalties for exceeding air and water quality standards, and payments for waste depositions. The resources from the fund are used to provide financial support for investments and governmental programs for the environment. In the period 1995–2000, more than 40 percent of the total expenditures of the State Environmental Fund supported air protection projects (table 3.3).

Thus, shortly after the breakup of the communist system, the Czech Republic as part of Czechoslovakia and after 1992 as an independent country developed a strong system of air pollution control compatible with EU legislation and the provisions of protocols to the LRTAP convention. This early adjustment to international standards is surprising in view of the high cost for the electricity industry. Even for EU countries, it took years to reach an agreement with the energy sector about air pollution regulations.[27] The level of compliance with EU air pollution directives by some member states remained imperfect due to domestic opposition.[28] In the Czech Republic, there was little visible resistance on the part of the electricity industry to the adoption of the Act on Clean Air. Moreover, the industry achieved spectacular compliance with the strict provisions of the legislation. What accounts for this surprising environmental strategy of the Czech electricity sector?

The Electricity Industry and the Act on Clean Air

Electricity production in the Czech Republic is highly concentrated in the Czech Electricity Company (CEZ). During the 1980s, CEZ was owned by the socialist government and controlled the production, transmission, and distribution of electricity. In the process of post-communist reform, eight regional distribution companies were created, separating the distribution of electricity from its production. CEZ was included in the first wave of privatization, and in April 1992 it was transformed into a joint-stock company. At the end of the 1990s, the government's National Property Fund owned a substantial proportion of the company's assets (67.6 percent), while the majority of the remaining shares were publicly traded on the Czech stock exchange and owned by Czech as well as foreign investors.[29] CEZ accounted for approximately 75 percent of the electricity produced and supplied in the Czech Republic. It owned the majority of the coal-fired power stations, 13 hydroelectric plants, the Dukovany nuclear plant, and the Temelin nuclear plant. Smaller cogeneration units connected to towns and industrial facilities provide the rest of the power supply. In 1998, 69 percent of CEZ electricity was generated in coal fired plants, 28 percent in the Dikovany nuclear station, and 3 percent by hydropower stations.[30] CEZ was also the owner and operator of the national transmission grid. It sold electricity to the eight distribution companies. Only a small percentage (about 4–10 percent) of the electric energy produced by the company during the 1990s was exported, with the share of exports increasing toward the end of the decade.[31] CEZ electricity imports were realized mostly to the extent necessary to ensure the reliable operation of the power system.[32]

In 1990–1991, when the Czechoslovak Act on Clean Air was formulated, CEZ had little economic interest in supporting or complying with strict environmental standards. Most of the earlier accounts on the Act on Clean Air, including comments by government officials, emphasized that the provisions of this legislation would be extremely costly for the sector. Estimates made by CEZ in 1991 were in the range of 68 billion Czech Koruna (US$2.3 billion) to be invested in desulfurization equipment alone.[33] The economic strain was compounded by the short implementation period. The task was indeed so costly that many analysts doubted the ability of the industry to comply with the new air pollution legislation.[34]

Unlike the chemical industry which could reasonably expect that the cost of stricter chemical safety standards would be offset by improved access and performance in EU markets, the electricity industry could not rely, at least during the first decade of post-communist transformation, on any significant benefits from trade to compensate for the high expenditures associated with tough air emission standards. CEZ was not involved extensively in international trade throughout most of the 1990s. Immediately after the collapse of the communist system, the Czechoslovak government emphasized the need for self-sufficiency and security of energy supply, and did not encourage exports of electricity in the early 1990s.[35] Moreover, in the early 1990s, when the Czechoslovak Act on Clean Air was adopted, the electricity markets of West European states were highly protected and efforts to gradually liberalize the internal energy market within the EU were in their incipient stage.

In 1991, the Czech electricity sector was thus confronted with air pollution legislation that was going to impose significant costs without bringing, at least in the foreseeable future, substantial benefits. The narrow economic payoffs associated with the clean air legislation dictated a basic interest against such policies. Similar to utilities in Poland and Bulgaria, the electricity sector in the Czech Republic had a strong preference against the imposition of stringent air emission standards. The Czech power sector, however, had limited political channels to effectively veto such legislation. The early move of the Czech Environmental Ministry and the Federal Committee of the Environment in preparing the air pollution law capitalized on the overwhelming political support for environmental reform to counter any possibility of an industrial veto. The narrow policy preferences of electricity utilities against costly emission standards could not effectively derail the reform of air pollution legislation, which was a central element of the program of the governing coalition.[36]

The ability of the power industry to lobby the Parliament against the Act on Clean Air and its provisions on air emissions was even more limited than in the executive branch. The position of parliamentary representatives on environmental reforms was often more radical than that of the national or federal governments. There was a strong anti-coal feeling in parliament, especially among representatives of highly polluted areas. The former head of the environment department of CEZ summarized the political dead-end the industry faced: "The power sector tried to resist such stringent regulations without much success. The public mood, reflected in the preferences of parliament members, was very much in favor of strict air-emission standards. Power plants were seen as the main culprits."[37]

Given the constraints of the Czech political context, an industrial strategy of outright resistance to the new air emission legislation was largely futile. Instead, CEZ focused on linking environmental cleanup to other policies that were going to influence the development of the sector, most notably issues of energy sector restructuring and access to investment financing. This bargaining approach was facilitated by the fact that governmental discussions of the clean air legislation proceeded simultaneously with efforts to restructure the country's energy industry. CEZ was able to maintain strong and direct ties to the coal and nuclear lobbies and to present a common position with respect to energy sector restructuring.[38] From this position, the electricity industry embarked on a complex negotiation strategy. The high cost of air pollution regulations became a bargaining chip that could be used to get concessions and favorable policy outcomes on important structural issues such as maintaining the integrated structure and dominant position of the company, improved access to financing for environmental investment, and the completion of the Temelin nuclear plant.[39]

A central concern for CEZ was its ability to maintain a highly integrated structure and a dominant position on the Czech electricity market. Early in the transition period, there were governmental plans and strong advice on the part of international agencies to de-monopolize the Czech and Slovak electricity sectors.[40] However, such reforms did not materialize. During the privatization of the Czech electricity industry, the bulk of production capacity remained under the control of CEZ. Only one power plant, Opatovice, was privatized as an independent power producer. CEZ used, along with other measures, the investment requirements of the Act on Clean Air as a rationale to keep CEZ vertically integrated. The company advanced similar arguments to resist quite effectively any attempts on the part of the government to increase competition in the supply of electricity throughout the 1990s.[41] Provisions for the liberalization of the Czech electricity market were adopted only by the 2000 Energy Act, which entered into force in 2001 and conforms to the EU electricity directive of 1996 (EC/96/92). The 2000 Energy Act includes provisions such as gradual liberalization of access to the electricity networks in 2002 and 2003, creation of an independent regulator, and a licensing system affecting the electricity market.

Another structural issue closely linked to the ability of CEZ to meet the requirements of the Act on Clean Air was the increase of nuclear capacity for electricity production. The construction of a second nuclear plant in the Czech Republic had started during the communist period under Soviet design, and its completion after the democratic changes was a highly contested issue. Environmental groups opposed it strongly and demanded that the nuclear facility be shut down. Neighboring Austria, which has no nuclear facilities of its own, actively opposed the construction and exploitation of the Temelin plant, situated only 50 miles from its border.[42] The first environmental program of the Czech Republic also advised against the further use of nuclear energy, emphasizing that "the substitution of fossil fuel with nuclear energy, considered a basic strategic orientation in Czechoslovakia in the past, is incompatible with an ecological and safety point of view."[43]

Despite strong pressure from environmentalists and neighboring countries, the completion of two nuclear reactors in Temelin that would compensate for the closing down of the oldest coal-fired plants was an important element of the strategy of the electricity sector. The expansion of nuclear power capacity was also in line with the government's concern with the security of the electricity supply and with maintaining reliance on domestic energy sources.

The emphasis on the supply of electricity, combined with the use of end-of-pipe technologies to achieve lower pollution levels, has been widely criticized by the environmental community in the Czech Republic. A great number of non-governmental organizations in the Czech Republic are active in the area of energy policy and conservation, among them Ekowatt, the Program for Energy Efficiency and the Slunicko Foundation, SEVEN, Greenpeace Praha, the Rainbow Movement, Children of the Earth, Calla, and Liga Energetickyh Alternativ. These organizations have continually demanded during the 1990s greater attention to efficiency improvements, deregulation of the electricity supply, and allowing third-party access to the national grid to increase the choice for more efficient and environmentally friendly sources of energy. Environmental organizations point out that there has not been a concerted effort on the part of the government to create economic incentives for end users to conserve energy as a way to reduce the environmental impact of the sector. Paradoxically, the emphasis on increased investments in production capacity and coal desulfurization may even decrease the total efficiency of energy production since desulfurization technologies tend to be energy intensive.[44] The point of view of the energy conservation community with respect to the government's approach to energy and environmental issues was summarized eloquently by one representative of an environmental organization:

> The government is still focusing mainly on the supply side. One of the big mistakes was the decision to continue the construction of the Temelin nuclear plant, instead of investing in retrofitting power plants and in energy efficiency programs. Most environmental investments in the energy sector were made in desulfurization equipment. Deeper concern with energy efficiency is still missing.[45]

From the perspective of elected officials and of the energy sector, however, policies that relied on expanding the share of nuclear generation and using end-of-pipe technology to reduce emissions from coal-fired power stations were politically efficient. These policies were in line with the goal of the environmental administration to achieve fast and visible reduction of air pollution. The use of end-of-pipe sulfur scrubbing technologies, although potentially very expensive, served the objective of improving air quality while maintaining reliance on domestic coal. Finally, the expansion of the share of nuclear energy corresponded to the interest of the electricity sector to strengthen its production capacity on the basis of domestic resources that were considerably less expensive than imported oil and natural gas. The decision to complete the nuclear power plant also increased the long-term prospects for surplus capacity in Czech electricity production and thus the possibility for a future increase in electricity exports.

As a result of the ability of CEZ to link environmental policies to sector restructuring and market position, the company was able to take a long-term view with respect to the requirements of the Act on Clean Air. Environmental cleanup became a part of a broader and more ambitious development strategy. In the shadow of future regulations, there were certain advantages for CEZ to undertake the necessary environmental improvements early in the transition period, when it was able to exploit its dominant position on the market and the backing of governmental support. From this perspective, the most important issue was not the magnitude of the environmental costs, but the ability of the company to find affordable financing to cover these costs.

The access to investment financing was another central question raised by the Czech electricity industry in relation to the Act on Clean Air. In the early 1990s, there was a great deal of uncertainty about the future revenues of CEZ and the level of funding available from domestic sources. The resources of the company depended on domestic demand, which was expected to fall as a result of the economic contraction, on electricity and fuel prices, on the availability of government subsidies, and on the extent to which electricity export restrictions were removed.[46] The lobbying strategy of CEZ included all possible aspects of access to resources for environmental improvements—from governmental subsides, to electricity pricing and international loans.[47]

CEZ was able to obtain preferential financing from the State Environmental Fund. Equally importantly, CEZ received governmental backing in its efforts to raise investment capital internationally. The World Bank loan for Energy and the Environment, negotiated with the Czechoslovak government in 1991, provided an important input in the financing of clean air improvements early in the transition period. CEZ was the main beneficiary of the loan, which allocated resources for power plant modernization, for the installation of desulfurization equipment at the Prunerov II plant, as well as for improvements and equipment for dust collection at the worst polluting CEZ coal-fired stations.[48] The loan covered only a small part of the investments needed to achieve the necessary environmental improvements. However, as industry sources underscore, CEZ "benefited greatly from a World Bank loan granted to the company as the first ever non-government Eastern Bloc entity, and subsequently from favorable credit ratings awarded to CEZ by leading rating agencies."[49] The World Bank's trust in the company, as well as the willingness of the Czech government to guarantee commercial loans for CEZ, increased creditors' confidence and greatly enhanced its ability to receive financing from international markets. As a consequence, CEZ was able to undertake throughout the 1990s an ambitious environmental program, supported by its own revenues, as well as by commercial loans and international assistance from the World Bank, the European Investment Bank, and the PHARE Black Triangle Program.[50]

CEZ formulated its environmental program in 1992, largely on the basis of the Act on Clean Air and the anticipated completion of the Temelin nuclear power station. The program set the ambitious objectives to reduce SO_2 emissions by 90 percent, flue ash emissions by 90 percent, and NOx emissions by 60 percent until 2000.[51] The realization of the environmental program of CEZ involved installation of desulfurization equipment in coal fired power stations, the introduction and renewal of electrostatic precipitators, technological measures to reduce emissions of nitrogen oxides, construction of fluidized-bed boilers at a number of power stations, as well as the gradual phasing out of older coal-fired plants. The installation of desulfurization technology was the most expensive environmental investment, with total cost of CZK 26.9 billion (approximately US$0.9 billion).[52] According to 1998 estimates, in the period 1993–98, the company put 25 percent of all its investments into the construction of desulfurization equipment and fluidized-bed boilers.

Another 35 percent of all investments in that period supported the construction of the Temelin nuclear plant.[53]

By 2000, CEZ had achieved a remarkable degree of reduction of acidifying air emissions. All coal-fired power stations in operation fulfilled the conditions specified in Czech clean air regulations. As part of the phase-out program, a total of 1,965 MW in the form of generation units and boilers were gradually shut down from 1991 to 1999. As a result, the contribution of CEZ to air pollution was reduced dramatically. Between 1992 and 2000, CEZ reduced its dust emissions by 96 percent, its SO_2 emissions by 90 percent and its NOx emissions by 51 percent (table 3.4, figure 3.1). The fossil fuel desulfurization program of CEZ is widely viewed as a big success story for the company, often described as one of the most ambitious environmental programs undertaken by an electricity utility in Europe.[54] By 1999, the ratios of SO_2 and NOx emissions per unit of thermal power produced were lower in the Czech Republic than the average for other OECD countries.[55]

As the prospects of the partial opening of West European electricity markets increased in the second half of the 1990s with the adoption of the 1996 EU directive on the internal electricity market, the interest of CEZ in future participation in European markets reinforced its strategy of improved environmental performance and increased nuclear capacity.[56] As the argument developed in this book would predict, the new opportunities for closer involvement in the European energy market affected the environmental position of the Czech electricity industry. The anticipated completion of the Temelin nuclear plant promised sufficient production capacity of CEZ to participate in regional markets once they were to liberalize. As part of closer energy cooperation in Europe during the second half of the 1990s, the management of CEZ worked actively to create conditions for parallel interconnection of the Czech power system with the Union for the Coordination of Production and Transmission of Electricity (UCPTE). At the policy level, CEZ provided strong support for a new energy strategy of the Czech Republic that would achieve gradual adaptation to the requirements of the EU electricity directive. The company itself also took steps to adjust to the standards of the emerging internal electricity market of the EU. In 1998, for example, the Board of Directors of CEZ established a subsidiary

Table 3.4
Annual emissions of SO_2, NOx, and solid particles from CEZ power stations (thousand tons/year). Source: CEZ 2002.

	1992	1993	1994	1995	1996	1997	1998	1999	2000	Reduction 1992–2000
SO_2	769.1	719.1	644.8	609.5	481.2	310.0	159.6	63.5	73.2	90%
NOx	128.5	122.2	77.4	75.3	71.0	67.4	56.9	52.7	62.9	51%
Solid particles	57.9	55.4	17.7	11.5	11.4	10.6	7.0	2.5	2.3	96%

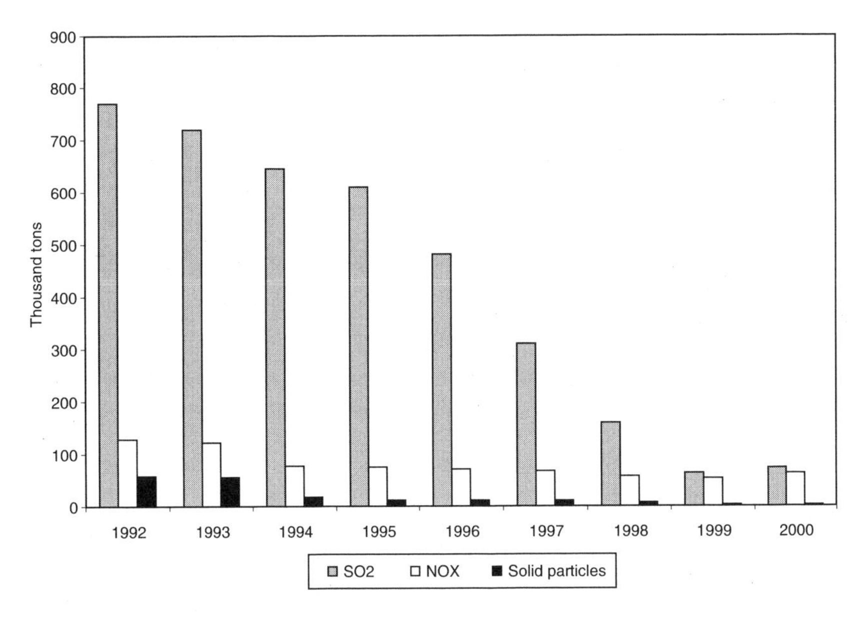

Figure 3.1
Annual emissions of SO_2, NOx, and solid particles from CEZ power stations (thousand tons/year). Source: CEZ 2002.

company to handle the operation of the transmission system, thus meeting the requirement for the unbundling of accounts set by the EU Electricity Directive. The ability of CEZ to comply with air emission standards compatible with those of the EU reinforced its new, long-term strategy of international market integration. The opening statement by Chairman of CEZ in the 1998 Annual Report speaks directly to this effect:

> During the years 1993–1998, CEZ completed one of the most demanding environmental programs in the Czech Republic, at the cost of CZK 45 billion. All power stations owned by CEZ were desulfurized and brought into compliance with the Act on Clean Air. In terms of the short completion time and the sheer volume of expenditure, this capital investment project is unparalleled not only in Central Europe, but in the world as well. The project puts the energy sources of CEZ in the same starting position as any plant in Western Europe and gives us the opportunity to participate in Western European markets.[57]

Thus, shortly after the democratic changes, the Czech electricity sector undertook significant measures to reduce air pollutants in accordance with Czech and European standards. This compliance strategy was, on one hand, dictated by strong domestic legislation adopted against the short-term interests of the industry. On the other hand, the compliance of CEZ with strict air emission standards was facilitated by a compensatory policy bargain, which allowed the sector to maintain its monopoly in electricity supply to the Czech market, to increase substantially its nuclear capacity, and to gain access to investment financing. The later orientation of CEZ toward participation in the slowly liberalizing European energy markets reinforced the importance of improved environmental performance. The compliance of the sector, in turn, had profound implications for the implementation of air pollution policies in the Czech Republic.

Compliance with International Standards

In the early transition period, the Czech government made a strong commitment to reform its air pollution legislation in accordance with European standards. By the end of 1998, when the Czech Republic had started official accession negotiations with the EU, it had achieved a high level of harmonization with EU air emission standards and with the provisions of the Second Sulfur Protocol to the LRTAP convention. The 1997 "Environmental Legislative Gap Analysis for the Czech Republic," prepared for the Ministry of the Environment and for the Environment Directorate General of the European Commission, concluded that Czech legislation was 100 percent aligned with the provisions of the Large Combustion Plant Directive. Czech ambient air quality standards were also closely compatible with the requirements of EU directives.[58] After the publication of the report, the Czech Ministry of the Environment undertook further measures to harmonize technical provisions related to ambient air quality assessment and the management of zones with high air pollution levels. Thus, within the first decade of post-communist transition, the Czech Republic attained a high level of harmonization with the EU Acquis related to industrial emissions into the air.

In the period 1990–2000, the country also achieved a remarkable degree of implementation of its strict air pollution legislation. According to data from the Czech Ministry of the Environment, in the period from 1990 to 2000 the emissions of SO_2 decreased by 86 percent, the emissions of solid particles by 91 percent, and of NOx by 46 percent (table 3.5, figure 3.2). The downward trend of air pollution emissions was maintained throughout the transition period despite the renewal of economic growth in the country after 1992. The high concentration of sulfur dioxide and particulate matter into the air is no longer considered the most urgent problem of Czech environmental policy. The focus of Czech efforts to improve air quality has now shifted to the management of emissions from mobile sources, whose number has increased dramatically in the post-communist period.[59]

Over the course of the 1990s, the Czech contribution to transboundary air pollution decreased dramatically. By 2000, the country had overcomplied by a large margin with the 1994 Second Sulfur Protocol to the LRTAP convention. In the 1999 negotiations of the new multi-pollutant protocol of the convention, on the basis of already-adopted domestic legislation the Czech Republic committed to some of the highest reduction targets specified by the protocol, undertaking to reduce SO_2 emissions by 85 percent, NOx emissions by 61 percent, ammonia emissions by 35 percent, and VOC emissions by 49 percent of 1990 levels by 2010. In the 10 years after the velvet revolution, air pollution policies and the international environmental image of the Czech Republic changed dramatically. Though still one of the most industrialized Central and East European states, the Czech Republic is no longer considered one of the biggest polluters in Europe.

Conclusion

The Czech air pollution reforms can be characterized as anticipatory adaptation to European norms. The early implementation of strict air emission standards in the Czech Republic may seem surprising given the high cost implications of international air pollution regulations for the electricity industry and the uneven compliance record of wealthier West European states. In the early 1990s, when the Czech air pollution legislation was formulated, the Czech power generation industry, characterized with a predominant orientation to domestic markets, anticipated

Table 3.5
Annual emissions of SO_2, NOx, and solid particles in Czech Republic, 1990–2000 (thousand tons/year). Source: Ministry of the Environment of the Czech Republic 1999, 2000, and 2001.

	1990	1991	1992	1993	1994	1995	1996	1997	1998	1999	2000	Reduction 1900–2000	Reduction 1992–2000
SO_2	1,876	1,776	1,538	1,419	1,278	1,091	946	701	443	269	264	86%	83%
NOx	742	725	698	574	434	412	432	423	412	389	397	46%	43%
Solid particles	631	582	501	441	355	201	179	128	86	67	57	91%	89%

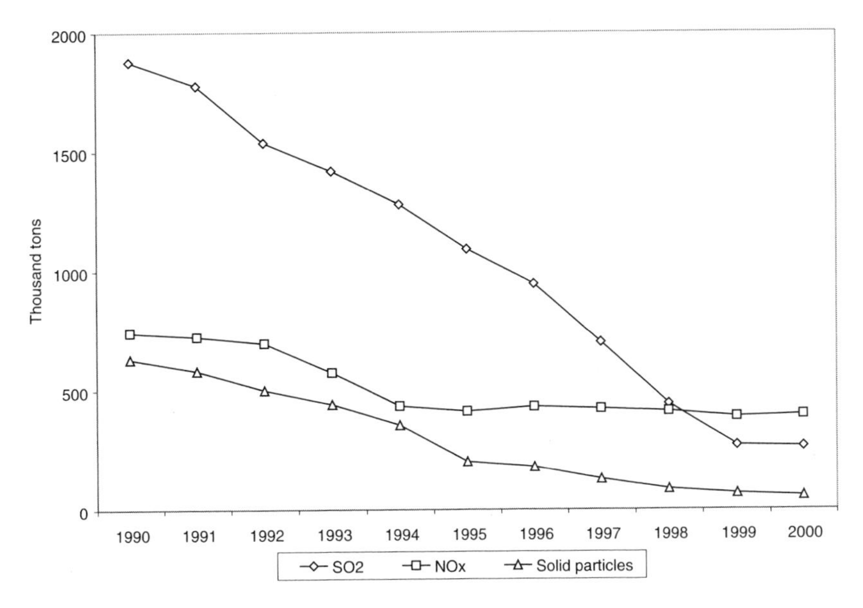

Figure 3.2
Annual emissions of SO_2, NOx, and solid particles in Czech Republic, 1990–2000 (thousand tons/year). Sources: Ministry of the Environment of the Czech Republic 1999, 2000, 2001.

high costs of compliance and little immediate benefits from greater regional integration.

As evidenced by early policy statements and by interviews with representatives of the sector, the electricity industry in the Czech Republic had a strong interest against the adoption of strict air emission regulations. The environmental strategy of Czech electricity utilities, however, was quite different from that of the utilities in Bulgaria and Poland, which were able to resist and to delay the application of strict command-and-control regulations. The Czech power generation sector had few opportunities to oppose the comprehensive Act on Clean Air adopted by the Federal Parliament of Czechoslovakia in 1991. Shortly after the adoption of the air-protection law, the Czech Electricity Company developed and implemented a far-reaching emissions reduction program, achieving compliance with strict domestic and international standards by 1999.

In many ways the environmental strategy of the Czech electricity sector during the late 1990s seems akin to that of the chemical industry, described in Part I of the book. However, there is an important difference in the nature of the incentives that motivate these seemingly common positions. The improved environmental behavior of chemical companies and their support for the approximation of EU chemical safety standards was motivated by international market pressure and immediate economic incentives. The position of the Czech utilities, in contrast, was influenced more strongly by policies and political bargains, rather than by economic costs and benefits. For the Czech electricity sector, environmental cleanup was part of a complex policy deal, which would ensure its integrated structure, as well as increased production capacity and access to investment finance that facilitated a long-term orientation to European markets. It was, thus, a set of policies negotiated between the government and industry that stimulated the growing interest of the Czech Electricity Company in international markets as well as its high level of compliance with air pollution regulations. In an interesting analytic twist, the apparent similarity between the environmental strategies of the Czech chemical and electricity industries highlights the central difference between the politics of international regulatory harmonization in the areas of chemical safety and air pollution.

Unlike in the chemical safety cases, where strong positive incentives associated with integration dominated the policy process and overshadowed cross-national differences, in the case of air protection, domestic political structures played an important role in shaping national responses. One important aspect of the Czech policy-making process was the limited ability of the electricity sector to veto air pollution legislation. The early move of the Czechoslovak Ministry of the Environment in proposing new air pollution legislation, while environmental concern was high, and the important role of the parliament in passing comprehensive air pollution legislation limited the opportunities for vested interests in the Czech Republic to block the reform of air pollution regulations. The Czech case thus seems to confirm the proposition, advanced by comparative analyses of policy change, that greater number of veto points decreases the likelihood of reform in the presence of strong opposition.

It will be misleading, however, to present the success of the Czech air pollution policy, and especially the high rate of implementation of air emission standards, simply as a story of political coercion of industrial interests in the name of the public good. While the preemption of industrial opposition was important for the timely adoption of strict air pollution legislation, there was also a strong element of compensation in the politics of air pollution reform in the Czech Republic. The industry and the government engaged in a process of logrolling and policy linkage that facilitated the negotiation of a compensatory bargain. For the industry, the bargain included maintaining an integrated structure and a monopoly position on the Czech market throughout the 1990s, increased nuclear production capacity, and governmental support in raising investment finance internationally. The government also supported the environmental investment program of CEZ both through the State Fund for Environmental Management and facilitating access to international lending. The leading economic performance of the Czech Republic compared to other post-communist states through most of the 1990s further strengthened the capacity of the government to raise international financing and to maintain policy support for improved environmental performance and development of the Czech electricity sector. The compensatory bargain justified the high cost of environmental improvements for electricity utilities during the transition period.

As EU electricity markets started to open toward the end of the 1990s, the early politically determined strategy of environmental cleanup of the Czech electricity industry supported its increasing orientation to international markets, which as anticipated by this analysis is likely to increase direct transnational pressures for cleaner generation.

Industry-government relations and bargains represent an important element shaping the politics of air pollution regulations in the Czech Republic. What seems to be missing in the picture of Czech air pollution politics is a continued influence of the environmental movement. While environmental activists in high government positions put the reform of air pollution on the agenda in the early transition period, benefiting from the high public concern of that era, some of the outcomes of the reform strongly contradicted the interests of Czech environmental groups. Environmentalists supported the objective of clean air protection, but they fundamentally opposed the emphasis on nuclear power development and on supply side energy management. However, efforts to reverse these policies have faced multiple setbacks during most of the transition period when the public and the government were predominantly concerned with economic restructuring and growth.[60] The closed bargain between the industry and the government on nuclear power development and the lock-in of this bargain with high level of investment considerably limited the leverage of environmental advocates in the Czech Republic. In Poland and Bulgaria, environmental groups also played a relatively weak role in shaping air pollution regulations, although in both countries environmentalists were more successful in resisting nuclear development in the energy sector during the 1990s.

The case of Czech air pollution reform thus illuminates the significance of two domestic characteristics in facilitating the adoption of costly international standards: the number and type of institutional veto points in the policy-making process and domestic capacity for compensation. This case also demonstrates, however, that each of these characteristics in itself does not provide a sufficient explanation of the rapid policy reform and implementation. The weak veto of the Czech industrial opposition facilitated legislative changes, but their implementation could have been thwarted in the absence of an agreement on a strategy for implementation and compensation. On the other hand, even if opportunities for a compensatory bargain existed, the strategies of industrial actors with respect to environmental regulations could have been quite different if these actors held a strong veto in the policy-making process. The story of policy linkage and successful commitment to policy change on the part of the Czech government indicates that in order to understand cross-country differences in the adjustment to international air pollution standards, it is not sufficient to focus on discrete institutional variables. While it is useful from a theoretical perspective to specify the independent significance of institutional variables, the analysis of interesting policy puzzles often requires a synthesis of theoretical knowledge and the recognition of interactive effects of multiple political factors. With this insight in mind, the next two chapters turn to the cases of air pollution reform in Poland and Bulgaria.

4

Poland: The Bargain of the Electricity Industry

Air pollution was a pressing environmental problem in Poland at the end of the 1980s. The dramatic deterioration of air quality during the communist development of the country was a consequence of rapid industrialization, a disproportionate share of heavy industry, inefficient enterprises, and almost exclusive reliance on coal for the production of electricity and heating. Between 1987 and 1989, Poland emitted nearly 4 million tons of SO_2, 2.5 million tons of particulate matter, and 2.5 million tons of NOx per year. Concentrations of these pollutants in major metropolitan areas exceeded permissible standards by levels that had negative health and ecological impacts. Poland was the third largest source of sulfur emissions in Europe after the Soviet Union and the German Democratic Republic (accounting for about 10 percent of all SO_2 emissions)[1] and faced strong pressure in the aftermath of communism to conform to European air pollution standards.

The Polish approach to air pollution reforms differed from that of the Czech Republic and Bulgaria. The first post-communist government of Poland, as in the Czech Republic, acted quickly on the task of reforming the country's environmental protection system. But relative to the Czech Republic, Poland appeared much more concerned about the cost of air pollution regulations for industry, and it was unwilling to ratify protocols to the LRTAP convention and introduce costly EU standards in the early transition period. By the end of the 1990s, however, Poland had achieved gradual but consistent implementation of European norms despite its earlier unwillingness to commit to their adoption. In order to account for the remarkable evolution of Polish air protection policy, this chapter again focuses on industry-government relations and the role of domestic institutions in facilitating costly policy reforms.

The Political Context of Environmental Reforms

The end of the communist system in Poland was marked by the historic agreement between the communist government and the Solidarity opposition to hold round-table talks in the spring of 1989. The negotiations resulted in the first semi-free elections in communist Poland, in which Solidarity won all the parliamentary seats for which it was allowed to compete. The political victory of Solidarity signaled the beginning of the democratic transition in Poland and the downfall of communism throughout Eastern Europe. The first Solidarity government of Tadeusz Mazowiecki used the broad-based support for reforms as a "window of opportunity" to initiate rapid measures for economic stabilization and market orientation of the Polish economy, and for its integration in Western markets and institutions.[2] After an initial period of hyperinflation, the rate of inflation stabilized, market shortages disappeared, the budget deficit was eliminated and the trade balance improved. After 1992, growth of the gross domestic product (GDP) increased rapidly, reading 7 percent in 1995. In 1996, Poland became a member of the OECD. (See table 4.1.)

The early post-communist period presented opportunities to advance the objectives of environmental protection in Poland. The Solidarity round table included a special subgroup on the environment, with strong representation of the independent ecological movement, which adopted guidelines and protocol on environmental reforms.[3] Compared to the Czech Republic and Bulgaria, where environmental activism also accompanied the break of the communist system, the independent ecological movement in communist Poland was characterized by greater organizational strength, expertise, and diversity, which were supported during the 1980s by strong opposition networks and the Catholic Church. The nuclear accident at Chernobyl in 1986 and the visible peace component within the movement also contributed to a strong anti-nuclear sentiment in the public mind.[4]

Table 4.1
Economic indicators for Poland. Source: World Bank 2002.

	1990	1991	1992	1993	1994	1995	1996	1997	1998	1999	2000
GDP growth (%)	—	–7.0	2.6	3.8	5.2	7.0	6.0	6.8	4.8	4.1	4.0
GDP per capita (1995 US$)	2,990	2,772	2,835	2,936	3,081	3,293	3,488	3,722	3,899	4,061	4,223
Inflation (%)	555	77	45	37	33	28	20	15	12	7	10
Unemployment (%)	6.5	11.8	13.3	14	14.4	13.3	12.4	11.2	10.7	12.5	16.7

As a result of the active participation of ecological organizations in the anti-communist opposition and in the round-table negotiations, the environment was a salient political issue in the early 1990s. A survey of the early 1990s, commissioned by the Institute for Sustainable Development, indicated that at that time 80 percent of respondents were highly or very highly concerned with the state of the environment and only 3 percent were not concerned at all. Moreover, the same research also indicated that 58 percent of respondents were willing to bear some social costs such as unemployment and shutdown of polluting enterprises for improved environmental conditions. The environment was the third most important issue in the Solidarity election campaign in 1989 and environmental reform was included among the priority tasks of the first post-communist government.[5] International institutions and Western allies of the Solidarity government also manifested support for environmental improvements in Poland. By the end of 1991, Poland had received US$215 million in foreign assistance for environmental projects, including aid from thirteen countries, the World Bank, and the European Community.[6]

Shortly after its inauguration, the first post-communist government adopted the National Environmental Program of Poland.[7] On the basis of the program, the government started far reaching reforms to strengthen the institutional and financial capacity for environmental management. This work was also supported by a World Bank loan of $18 million for environmental management in Poland, negotiated in 1990. The close cooperation with the World Bank, and the involvement of activists and environmental economists affected significantly the direction of the Polish environmental reform. The most distinctive elements of the new system for environmental management included decentralization of environmental protection, strengthening of monitoring and enforcement, and an emphasis on economic incentives and implementation capacity.[8]

After 1990, the departments of environmental protection at the regional level assumed a wide scope of competencies in the administration of environmental law.[9] These departments are entitled to determine regional charges, standards, and other environmental requirements for enterprises, to collect fees and inflict penalties, to dispose of part of target funds, and in some instances, to establish standards stricter than those in the rest of the country. This system, together with the publication of national and regional lists of the most polluting enterprises, increased local control and scrutiny over the activities of polluting entities, and the pressure for improved performance.[10]

The 1991 Law on the State Environmental Protection Inspectorates strengthened the enforcement capacity and the power of the national and regional Environmental Inspectorates, which became known in post-communist Poland as the "green police." This law separated regulatory functions from monitoring and enforcement responsibilities by making inspectorates independent of regional administrations and thus of local political interests. Regional and national inspectors have the power to monitor pollution, to impose noncompliance fines and penalties, and even to shut down enterprises endangering the environment.[11]

The government also strengthened the system of environmental user fees and fines as an economic mechanism for enforcement and revenue generation. The fees for emission of SO_2, for example, reached 85 euros per ton in 2000, among the highest in Europe.[12] The penalty for noncompliance with permitted standards for SO_2 and NOx emissions was about 850 euros per ton.[13] The environmental fees and penalties in Poland, as in most states in the world, did not reach the level of pollution-abatement cost so as to provide a sufficient economic incentive for pollution reduction. However, the substantial increase of fees and fines motivated big polluters to invest in improved environmental management, and had an important revenue raising function.[14]

Poland also established in the early transition period a system of funds for environmental protection, including the National Fund for Environmental Protection and Water Management, regional funds, and nearly 2,500 local funds. Revenues for the funds are generated from fees for the emission of pollutants, discharge of wastewater, and dumping of wastes, and from fines for excessive pollution. Of these resources, approximately 10 percent remain in local funds, 54 percent are directed to regional funds, and some 36 percent go to the national fund.[15] The collected environmental fees and fines are recirculated into environmental investments through subsidies and preferential lending from the funds, which are often extended in conjunction with loans from the Bank

Table 4.2
Sources of environmental investments in Poland (percentages). Source: Council of Ministers of the Republic of Poland 2000.

	1991	1992	1993	1994	1995	1996	1997
Environmental funds	40	58	47	41	40	34	30
Investors' funds/loans	30	20	25	31	32	38	40
Central budget	5	5	7	5	5	5	3
Local budgets	20	13	16	19	18	19	23
Foreign assistance	5	4	5	4	5	4	4

for Environmental Protection. In 1992, the government also established the EcoFund, which manages Poland's debt-for-nature swap.[16] Ecological funds have played an important role in environmental protection in Poland, accounting for approximately 40 percent of environmental spending,[17] a share that decreased toward the end of the 1990s as a result of greater availability and role of commercial lending (table 4.2). Following the Polish model, similar structures were developed throughout Central and Eastern Europe, although with differing influence and capacity across states. The Polish environmental funds are among the best-managed and most politically influential systems of environmental financing in the region.[18]

The Polish environmental reforms of the early 1990s thus locked in capacity for environmental management and put an emphasis on priority setting, economic incentives, and institutional reform, rather than on command-and-control instruments and extensive legislative changes. The government established strong environmental institutions, which, according to many observers, could not have been created later in the transition period when public concern diminished. One official in the Ministry of the Environment summarized the significance of these structural changes as follows:

> We established two very good institutions—the environmental funds and strong inspectorates—very early on, when industries were not used to lobbying against the government, and when there was a lot of enthusiasm. The strength of these institutions allows for relatively good environmental management even now when enthusiasm has waned and there is much more opposition and opportunity to block reforms. The creation of these structures now would probably be impossible.[19]

Indeed, shortly after the initial cresting of public and governmental concern, the urgency of environmental issues fell out of the public eye. By 1991, the early political consensus on the need for economic and political reforms gave way to party fragmentation, power struggles among branches of the government, and instability of governing coalitions. Prolonged confrontation between the presidency, the parliament, and successive governments undermined the speed of the constitutional reform and the restructuring of the economy.[20] The major political parties had little interest in environmental issues after the initial period of high concern as evidenced by the absence of environmental debates in political campaigns and governmental programs after 1992, as well as by the weak position of the Ministry of the Environment in the administration.[21] As in other transition countries, the political presence of the environmental movement also experienced a slump after the early transition period. Despite the growth in the number of organizations after 1990, the majority of green groups remained active primarily at the local level. The movement was fragmented politically and lacked channels of access to political parties and governmental bodies. Relations with the Ministry of the Environment became tenuous and often confrontational.[22]

The governmental instability in Poland, the weak position of the Ministry of the Environment, and the low influence of environmental advocates caused significant delays in the reform of environmental legislation. The bulk of new environmental regulations were adopted on the basis of the provisions of the 1980 framework Act on Environmental Protection.[23] The 1980 act was considered outdated and inadequate for the modern requirements of environmental legislation,[24] but continued to stay in force after some amendments. A new Act on Environmental Protection was adopted in 2001. Despite the significant delays in the legislative program of the Ministry of the Environment, however, thanks to the institutional changes undertaken in the early transition period, Poland maintained a course of continued improvement in the state of the environment. This was in part a consequence of the restructuring and recession in the economy, but also as a result of better enforcement and increasing environmental investment. Environmental expenditures as a percent of GDP rose from only 0.7 percent of GDP in 1990[25] to 1 percent in 1991 and 1.6 percent in 1997, levels comparable to expenditures in OECD countries.[26]

As preparations for EU accession negotiations intensified in the second half of the 1990s, the environmental-policy agenda was reinvigorated both by the effort of the government to advance the adoption of the EU Acquis[27] and by the more active and instrumental approach of some environmental groups to politics.[28] In the parliamentary elections of 1997, the Freedom Union, one of the main political parties, ran its election campaign with a strong environmental program formulated by its Forum of Ecological Leaders. The presence of an ecological fraction within the Freedom Union signaled both the politicization of parts of the environmental movement as well as greater interest among the main political parties in environmental issues as a consequence of EU pressures.[29] Air pollution reform, as other environmental issues, was affected by the pull of international commitments, but remained a strongly contested issue in domestic politics and proceeded in close consideration of domestic interests and the position of the electricity sector in particular.

Air Protection in the Early 1990s

The system of air protection in Poland was first reformed as part of the comprehensive environmental program of the Solidarity government. The National Environmental Program of 1991 put a strong emphasis on reducing emissions of air pollutants. Five of the ten short-term priorities outlined in the document included air protection measures such as coal quality improvement, reduced emissions from transportation and industrial plants, improved monitoring, and compliance with ambient standards. The medium-term objectives of the national program required further improvements in air quality and the reduction of transboundary flows of harmful substances in the atmosphere.[30]

The priority accorded to air protection in the aftermath of communism was a reflection of the dramatic and visible deterioration of air quality in the country. During the 1980s, Poland's air, especially in close proximity to large industrial sites, has been consistently characterized as some of the most polluted in Europe. Compared to West European countries, Poland had several times higher emissions per capita and per GDP.

As in other communist states, this situation was a result of rapid and inefficient growth of heavy industry, low energy efficiency, strong reliance on coal for energy generation and heating, and lack of cleaning installations in industrial facilities.[31]

The air pollution problem was aggravated by the uneven spatial distribution of mining and industrial production. As numerous maps of the distribution of air emissions show, there is considerable concentration of pollution sources in selected regions of Poland, and close to large urban agglomerations. Six of the 49 administrative regions (wojewodztwa) accounted for about half of all air pollution emissions. The conditions were worst in the Katowice region. Despite having only 2 percent of the territory of Poland, this region accounted for 20–25 percent of national emissions of SO_2, NOx, and dust.[32]

High levels of pollution in the hot spots of Poland have been shown to correlate to increased occurrence of respiratory diseases, higher rate of infant mortality, and lower average life expectancy. A report by the World Bank presents data on the correlation between high concentrations of air pollutants and respiratory disease among children and adults, as well as evidence of a connection between air pollution in certain regions in Poland and excess infant and lung cancer mortality.[33] In addition to its adverse health impacts, air pollution in Poland caused high acidity of rain in the southwestern part of the country, resulting in forest loss, acidification of lakes and soils, and the corrosion of buildings and equipment. Economic evaluations estimated the cost of biological, social, health, and material damage from air pollution, indicating losses of up to 25.8 percent of national income and up to 35 percent of the income of the most polluted provinces for 1985.[34]

Air pollution in Poland was also a problem of international importance. The participation of the country in the LRTAP convention and its protocols, together with other big polluters on European scale such as Germany, the Former Soviet Union, and the UK, was important for pan-European efforts to control acidification. The ambition of Poland to join the EU further implied a strong obligation to adopt and comply with European norms. However, Poland was openly concerned with the cost implications of European regulations and unwilling to undertake commitments that would impose a significant burden on its industry. In 1985,

Poland was the only Warsaw Pact country that did not sign the First Sulfur Protocol, which mandated a uniform 30 percent reduction from 1980 emission levels for all parties to the LRTAP convention. Warsaw Pact states were generally willing to accept the requirements of the LRTAP protocols to support the strategic objective of East-West détente, but with little intention to make serious efforts at compliance. Poland was the only communist state that strayed from this course in the 1980s. In a study of the LRTAP regime, Marc Levy summarizes the reasons for Polish defiance: "Ironically, Poland was forced to remain outside the Sulfur Protocol because it had the most serious commitment to reducing emissions in the East Bloc. Its environment officials had set ambitious goals and studied implementation options; they knew that 30 percent reductions were not possible and were unwilling to make a promise they could not keep."[35]

The government's concern with the ability of Polish industry to bear the cost of international commitments and domestic regulations was reinforced by the direction of the post-communist environmental reforms, which emphasized capacity building, cost minimization, and incentives for implementation. This policy style is quite different, for example, from the Czech emphasis on command and control instruments to achieve fast and visible reduction of pollution. The Polish air pollution policy of the early 1990s put less emphasis on the adoption of international environmental standards, and greater emphasis on establishing enforceable domestic standards and incentives for compliance. In the period 1990–1992, the government increased the price of coal by more than 200 percent in dollar terms, tripled the electricity prices for industry, raised substantially the price of residential heating and electricity tariffs, and even considered a tradable permits scheme for acidifying emissions.

In 1990, the Ministry of the Environment issued an Ordinance on the Protection of Air, based on the provisions of the 1980 Act on Environmental Protection. The ordinance was strongly influenced by domestic concern with air quality, and commanded the approval of other important ministries, most notably the Ministry of Industry and Trade and the Ministry of Transport. It established ambient air quality standards, some of which were stricter than equivalent EU standards. The ordinance also introduced for the first time in the Polish regulatory framework nationwide emission limits for SO_2, NOx, and dust from combustion sources with capacity bigger than 0.2 MW, distinguishing between existing and new sources.[36] The limits vary according to fuel type and combustion process, and are set in terms of pollutants emitted per unit of fuel consumed, a measure that is not easily comparable to EU standards which set emission levels in terms of milligrams per normal cubic meter (mg/Nm^3).[37]

Approximate comparisons between the Ordinance on Clean Air adopted by Poland in 1990 and EU regulations and the Second Sulfur Protocol to the convention on LRTAP indicate some important differences between Polish and international regulations. The Polish standards of the early 1990s tended to be stricter for smaller combustion sources. However, they were more lenient than European emission limits for dust and SO_2 from combustion sources with capacity of 500 MW and over.[38] This is an important difference since the major proportion of power in Poland is produced in plants with capacity greater than 500 MW. The 1990 ordinance also did not impose a requirement for the use of best available technology or for specific rates of desulfurization, allowing greater flexibility in meeting emission reduction standards than EU regulations and the LRTAP convention.

The process of EU accession preparations intensified the pressure for further alignment of Polish air emission standards with EU and LRTAP regulations.[39] While compliance with international norms in the field of air pollution was seen as important for EU accession and for improving the environmental image of the country, any discussion of achieving EU standards inevitably evoked the question of the high cost of implementation. The reduction of sulfur emissions, in particular, emerged as one of the most sensitive problems on the Polish approximation agenda.[40] The Second Sulfur Protocol mandated not only significant SO_2 emission reductions. More problematic from the perspective of industry was the fact that both the EU Large Combustion Plant Directive as well as the Second Sulfur Protocol imposed strict emission and technology standards on individual combustion sources and required the application of costly equipment to new and rehabilitated power plants. The constellation of domestic industrial interests played an influential role in shaping Poland's strategy for adjustment to EU and LRTAP norms. The gradual reform of air pollution standards in the second half of the 1990s can be understood only in the context of the tough bargain between government and industry.

The Strategy of the Electricity Industry

Electricity generation together with district heating systems contribute the most significant share to air pollution in Poland. Power generation alone accounts for approximately 50 percent of SO_2 emissions, for a third of NOx emissions, and for 12 percent of dust emissions in the country.[41] Several waves of reorganization of the electricity sector during the early post-communist period resulted in a system that comprises three layers of companies: generation, transmission, and distribution. Big power plants with production capacity of 500 MW or greater predominate in this system, with 15 such stations accounting for more than 68 percent of installed capacity in 1990.[42] Four lignite mines, which supply thermal power plants, are considered part of the electricity sector. The Polish Power Grid Company (PPGC), established as a joint stock company in 1991 and owned by the State Treasury, is responsible for the transmission of electricity, for the management of the electricity grid, for conducting trade, and for ensuring the reliability and optimization of supply. It purchases power from individual suppliers and sells it to 33 regional electricity distribution companies, which also function as joint stock companies.[43] Although the sector as whole is under the direct supervision of the Ministry of the Economy, decisions related to production strategy, marketing, and investment are undertaken by individual power plants, which behave as independent decision makers.[44]

Poland updated its energy sector legislation in 1997 with the adoption of a new Energy Law, which established a framework for demonopolization and introduction of market mechanisms in energy management. The law envisaged the privatization of power companies, allowing competition in energy generation and supply, the elimination of state subsidies, and the introduction, in the longer run, of third party access to the national electricity grid. In 1998, the Economic Committee of the Council of Ministers adopted a program for the privatization of power and heating generation utilities until 2002. Through close consultation with trade unions, the government was able to reach an agreement about the necessity of structural changes and to work out guarantees related to employment, wages, social safety nets, and employee ownership of company stock.[45]

A close connection between electricity and coal production in Poland is a fundamental characteristic of the Polish power sector that has important implications for its ability meet strict air emission standards. According to 1989 statistics, more than 60 percent of electricity was generated from hard coal, 36 percent from lignite coal, some 3 percent from hydropower, and a small fraction from oil and gas.[46] In 1999, this production structure had not changed much: 92 percent of all electricity was generated in coal-fired power stations, 3 percent in hydro power stations, and 5 percent by industrial auto-producers, most of which used coal.[47] Because of the abundance of coal resources and their high social importance, the fuel base of electricity generation in Poland is unlikely to change radically in the near future. The predominance of coal in the Polish economy and in electricity generation is reinforced by the dependence of the country on Russia for imports of oil and gas and by strong public opposition to nuclear power. Unlike in Bulgaria and the Czech Republic, where nuclear stations contribute approximately 40 percent and 16 percent respectively to electricity generation, there were no plans for the development of nuclear capacity in Poland during the 1990s. The construction of a nuclear power plant of a Soviet type near Gdansk was halted immediately after the democratic changes, as a result of strong pressure from environmental groups and the mobilization of mass protests.[48]

The environmental consequences of the fuel structure of electricity generation in Poland are significant. Compliance with the LRTAP Second Sulfur Protocol and with the 1988 EU Large Combustion Plant directive necessitates costly investment in desulfurization equipment, switching to higher quality coal, and other modernization and technological measures. Official estimates of the cost of compliance with EU air emission standards in Poland range from US$1.5 billion to US$10 billion. The higher end of these estimates assumes a stricter interpretation of the EU Acquis and the Second Sulfur Protocol. Estimates of the cost of meeting the requirements of the Second Sulfur Protocol in the electricity sector alone range from US$1.8 to US$4.8 billion, depending on the flexibility given to power utilities for extended compliance periods and to use of cost-minimizing instruments rather than technology standards.[49]

The cost of strict international standards for environmental protection in electricity production could not be offset by significant benefits of integration as in the case of some export-oriented or multinational industries. Export of electricity represented only 8 percent of total generation in 1990 and 6.7 percent in 2000, while import of electricity accounted for 8 percent of total consumption in 1990 and 2.4 percent of total consumption in 2000.[50] European integration did not provide incentives to bear the increased cost of environmental improvement. Furthermore, as many industry sources pointed out, the technology and source-specific standards embedded in the Second Sulfur Protocol and the Large Combustion Plant Directive limit the flexibility with which electricity utilities can pursue environmental improvements.[51]

As predicted by the theoretical framework, and as can be expected from the narrow costs and benefits associated with the harmonization of international standards, the electricity industry in Poland voiced a strong preference against the adoption of costly international air emission requirements. A representative of the Energy and Environment Department of the Ministry of the Economy concisely stated the economic rationale behind this position: "The EU has very stringent regulations for the energy sector. These are too high for our industry, and will not be economically feasible to implement."[52] The Polish electricity sector has been able to effectively represent this position and to lobby for more gradual and flexible adjustment to international norms. Unlike in the Czech Republic, where the concerns about the cost of the Act on Clean Air for industrial actors did not present a political stumbling block for adopting strict air pollution regulations, in Poland the economic interests of the energy sector could not be ignored.

The power industry holds a strong veto in shaping air pollution regulations in Poland. The strategic position of the Polish electricity sector is based on three political and institutional characteristics: the backing of the highly unionized coal subsector and the considerable political influence of the energy lobby, the strong role of the executive branch in determining air pollution standards, and the emphasis on cost-minimization and implementation in the general direction of reforming environmental policy.

Power generation is closely associated with coal mining, which alone accounts for 1.5 percent of the employment in the country and for nearly 5 percent of its exports.[53] A significant share of coal production (up to 98 percent) is concentrated in the upper Silesian basin, where thousands of jobs depend on the future of coal mining. The electricity and coal industries in Poland are highly unionized. The Solidarity Union and the National Trade Union Accord are well connected to political parties: the former to the right of the political spectrum through post-Solidarity parties and electoral blocs, and the latter to the left of the political spectrum through the post-communist Left Democratic Alliance. As a consequence of the influential position of the energy lobby in society and in politics, successive governments proceeded with caution with restructuring and other policies pertaining to these industries.[54]

From its powerful political position, the electricity sector has been able to represent forcefully its concerns with respect to the cost of international air pollution standards. The strong influence of industrial interests has been further facilitated by the structure of the decision-making process and the leading role of the executive branch in the adoption of air emissions standards. In Poland, air pollution regulations are set by executive orders on the basis of the principles outlined in the 1980 Act on Environmental Protection, which was later replaced by the 2001 Act on Environmental Protection. The content and reform of air pollution regulations is thus highly dependent on inter-ministerial bargaining, where the narrow sector interests of individual ministries prevail more easily. This is different, for example, from the Czech approach in regulating air protection, which was based on a comprehensive Act on Clean Air adopted by parliament, where broader public concerns are better represented, and the opportunities for "industry capture" of specific regulatory provisions is more limited.

In the negotiation of ministerial ordinances for air quality and emission limits in Poland, the interests of the electricity sector are forcefully voiced by the Ministry of the Economy, by the PPGC, and by the

Chamber of Power Industry and Environment, created in 1993 to formulate the collective position of the sector. Executive plans have to be mutually agreed with the Ministry of the Economy, whose Energy and Environmental Protection Department is often directly involved in the negotiations of international protocols related to transboundary pollution and climate change. The differences of opinion are settled by consensus, giving significant leverage to industrial interests. As the Polish Chamber of Power Industry and Environment underlines in its publications, it is "difficult to overestimate the role of the sector in influencing new environmental regulations."[55]

The distinctive style of post-communist environmental management in Poland, which focused on institution building, cost-minimization, and implementation, further reinforced the necessity of taking close account of industrial capacity for compliance. Statements and publications of the Ministry of the Environment consistently stress the need to "optimize action" and to achieve reductions of air emissions in the most "cost-effective way." The close cooperation with the energy sector and the emphasis on cost minimization and implementation capacity is viewed as a way to demonstrate a credible commitment to compliance with international obligations.[56] This view is concisely summarized in the following statement of the Minister of the Environment: "Although in Poland the fuel and energy sector leads in the volume of pollutants emitted into the air—today it is, above all, a partner in those actions aimed at the improvement of the quality of the environment. The cooperation in the performance of programs and particular undertakings is a guarantee that the ecological requirements in Poland as well as international obligations will be met."[57]

Concern with the fundamental economic interests of the electricity sector thwarted the full adoption of EU and LRTAP air emission provisions during the early period of post-communist transition.[58] In the course of the 1990s, however, the environmental strategy of the electricity industry in Poland also evolved to reflect the constraints associated with the shadow of future EU integration and the domestic opportunities for partial compensation for the costs of environmental investment. The understanding that EU accession "is clearly a priority of the government, and that sooner or later compliance with EU standards will have to be achieved," motivated industrial representatives to get actively involved in shaping the ways in which EU approximation issues are addressed.[59] The management of the sector, as represented by the PPGC and the Ministry of the Economy, sought to maximize its influence over the content of new regulations. The strategy of the industry focused more and more not simply on blocking the adoption of costly international standards, but also on examining the options for the implementation of these regulations in the near or long-term future.

The PPGC invested resources in estimating the cost of compliance with EU directives and protocols under the LRTAP convention and participated in government and internationally sponsored analyses on the cost of EU law application. Such economic assessments sought to provide a basis for the position of the sector: "We have signed the Second Sulfur Protocol, but have not ratified it yet. The main studies are trying to answer the question can we ratify without excessive burden to the economy? How much would it cost?"—commented a representative of the PPGC.[60] On the basis of cost estimation and optimization analyses, the sector argued for a limited approximation strategy that allows greater flexibility than implied by the strict interpretation of EU directives and the Second Sulfur Protocol. This position was voiced by the PPGC, the Chamber of Commerce for Energy and Environmental Protection, as well as by representatives of individual power plants.[61] Thus in the course of the 1990s, the bargaining between industry and the government moved toward a more constructive dialogue. The question became not whether international norms would be achieved, but how and within what period of time.[62]

The evolution in the strategy of the electricity sector and its willingness to consider issues of international norm implementation was also facilitated by the availability of public support for environmental investments throughout the 1990s. In the period 1990–1998, power plants undertook investments in processes and technologies for the reduction of dust and gaseous emissions. Such projects included efficiency enhancing measures, the introduction of higher quality fuel in burning processes, and the construction of emission reduction equipment.[63] The strong capacity for extra budgetary environmental spending through the system of environmental funds greatly enhanced the ability of Polish

Table 4.3
Expenditures of the National Fund for Environmental Protection and Water Management of Poland (1995–2000, million US$). Source: Data in million PLZ, provided by REC 2001, converted into US$ by author using average annual exchange rates provided by the National Bank of Poland.

	1995	1996	1997	1998	1999	2000
Total	372.4	452.9	341.7	281.2	369.7	260.7
Air pollution	175.3	167.3	113.6	80.4	137.7	89.2
% air pollution	47	37	33	29	37	34

power plants to undertake emission reduction measures. Investments in pollution-abatement technologies in the electricity sector were supported by the national, regional, and local environmental funds, by the EcoFund, by international loans and grants, as well as by commercial loans and resources of the enterprises.[64] The annual investment in air protection in Poland in the period 1995–1997 constituted between 40 and 60 percent of the total environmental expenditures in the country.[65] The National Fund allotted increasing amounts of resources for air pollution abatement throughout the 1990s. The share of air pollution expenditures in the disbursement portfolio of the fund was between 29 percent and 47 percent in the period 1995–2000 (table 4.3).

The National Environmental Fund has participated in almost all desulfurization projects undertaken by power utilities. Despite the fact that the contribution of the Fund often amounted to only 10 percent of investment needs, these resources have had a catalytic effect in stimulating environmental investments and facilitating access to international and commercial financing. The public knowledge of the availability of financial resources for environmental improvements and the greater scrutiny of individual power plants by local regulators also increased the pressure on electricity utilities to reduce emissions and to seek cooperation with the government in setting new national standards.[66]

In conjunction with the decline in industrial production and the rationalization of production, targeted environmental investments contributed to the improved environmental performance of the electricity industry during the 1990s. In the period 1990–2000, the power sector reduced its emissions of SO_2 by 55 percent, its emissions of NOx by 55 percent, and its emissions of dust by 94 percent (table 4.4, figure 4.1). While the economic recession and the contraction in the demand for electricity in the beginning of the 1990s was the main cause of the initial sharp decline in the emissions of air pollutants, the trend of reduced emission levels was maintained even after the resumed growth of the economy and of electricity production in 1992. The measures for environmental improvements undertaken in the 1990s enabled many plants to meet domestic standards and increased the confidence both in the sector and in the government about the future capacity of the industry to comply with the basic requirements of international regulations.

The proactive position of the electricity sector, the close bargaining between the government and industry, and the ability to extend financial support for environmental investments facilitated the establishment of a compromise between environmental, foreign policy and industrial objectives in Poland. In September 1996, the Minister of the Environment, the Minister of Industry and Trade, and the Minister of Regional Planning and Construction agreed on a "National Program for Reduction of Emissions of Sulphur Dioxide by the Year 2010." The program outlines the general policies to be undertaken by industrial sectors toward achieving the goal of compliance with the targets for national emission reductions set by the Second Sulfur Protocol.[67] The Ministry of the Environment and the Ministry of Industry and Trade also adopted "The Program for the Reduction of Emissions of Sulfur Dioxide in the Power Supply Industry" as a measure for implementation of the national strategy for the reduction of SO_2 emissions by 2010. The program was based on analysis of alternative approaches for meeting international obligations, and was closely coordinated with the PPGC. It established the objective of reducing the SO_2 emissions of the electricity sector to 700 tons per year by 2010, which represented half of the country's emission ceiling set by the Second Sulfur Protocol. The document included a list of investments in desulfurization equipment to be undertaken by individual plants, and underlined the need for further changes in air pollution regulations.[68]

The agreement reflected in "The Program for the Reduction of Emissions of Sulfur Dioxide in the Power Supply Industry" was

Table 4.4
Emissions of SO_2, NOx, and dust for the electricity industry in Poland, 1990–2000 (thousand tons/year). Source: PPGC 2002.

	1990	1991	1992	1993	1994	1995	1996	1997	1998	1999	2000	Reduction 1990–2000	Reduction 1992–2000
SO_2	1,563	1,477	1,339	1,283	1,270	1,220	1,195	1,108	1,031	916	696	55%	48%
NOx	466	440	404	367	374	380	366	317	269	250	210	55%	48%
Dust	858	694	501	422	358	265	217	165	132	105	51	94%	90%

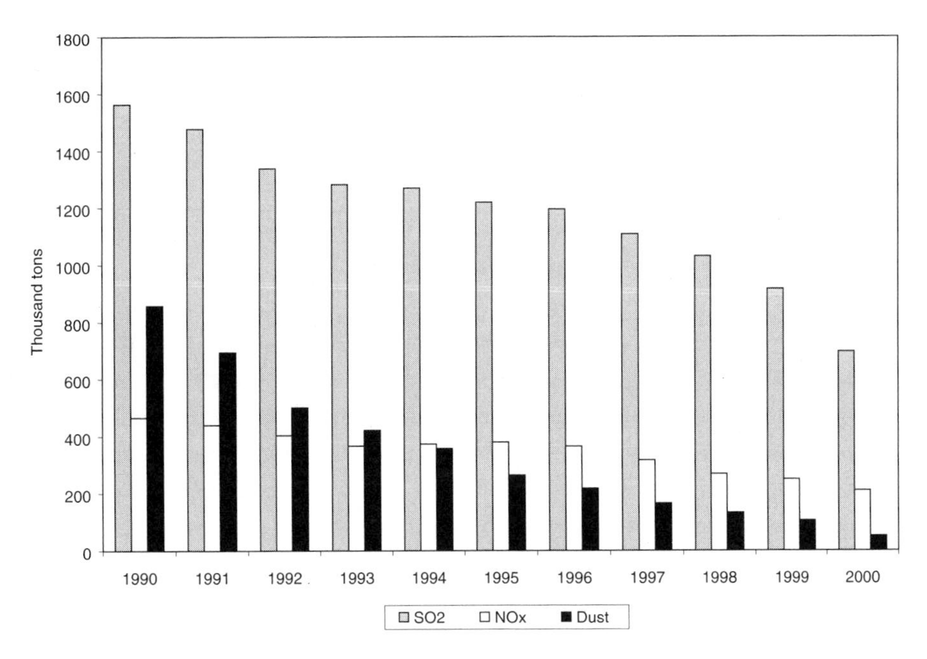

Figure 4.1
Emissions of SO_2, NO_x, and dust by the electricity industry in Poland, 1990–2000 (thousand tons/year). Source: PPGC 2002.

supported by specific financial guarantees for the electricity industry. These included a commitment on the part of the government to assist the sector in securing international loans and grants, and to recommend desulfurization projects as a financing priority in the operations of the National Fund for Environmental Protection and Water Management. The program was also backed by the establishment of long-term power purchasing contracts between utilities and the PPGC. In 1999, power purchase agreements covered more then 65 percent of electricity demand in Poland. While such contracts are criticized for creating price distortions that interfere with the objective of introducing market competition in the electricity sector, they guaranteed a relatively secure future and high financial returns for generating enterprises, which was viewed as essential for undertaking environmental measures.[69] The agreement between the Ministry of the Environment and the power sector on a long-term program for reducing SO_2 emissions reflects most clearly the evolution of the bargaining position of the industry under the impact of international constraints and domestic compensation. The position of this influential political actor, in turn, had a strong influence on the reform of air pollution policies in Poland and its ability to comply with EU standards.

Compliance with International Standards

Compared to the Czech Republic, which adopted even the strictest requirements of the 1988 Large Combustion Plant Directive in 1991, Poland followed a more gradual and limited approach to the adoption of international standards. A 1997 EU approximation report, prepared with the assistance of the European Commission, evaluated the air pollution legislation in Poland as significantly different and even "incompatible" with the EU Acquis. The report suggests that "the approximation of the Polish legal system of air protection against pollution to the requirements of EU law has to involve a reconstruction of the Polish system; it is not enough to just approximate the two types of standards or uniform measuring methods."[70]

By the second half of the 1990s, however, the Polish government had reached an agreement with industrial actors on the course and elements of further reforms to limit air pollution. On the basis of its 1996 agreement with the electricity industry, the Ministry of the Environment was able to make some important changes in the regulations concerning air pollution. In 1997, under the initiative of the ministry, the Parliament amended the section on Air in the 1980 Environmental Protection Act, despite delays in the formulation a new framework Act on the Environment. The amended Environmental Protection Act required the issuing of new ordinances on air pollution. In the spring of 1998, the Ministry of the Environment adopted the Ordinance on Admissible Concentrations of Pollutants into the Air and the Ordinance on the Emissions of Pollutants form Technological Processes and Technical Operations,[71] which together replace the 1990 Ordinance on Air Protection.

The 1998 air pollution regulations were based on an analysis of the EU Acquis, and aimed at as close an approximation of EU norms as was deemed feasible given the economic interests of important domestic industries and sought to reflect to the extent possible the 1996 agreements with the electricity sector. As one adviser to the Ministry of the Environment commented: "The new regulation on air emissions was prepared very intelligently. It approximates very closely EU directives but at the same time it phases in these regulations in such a way as to do as little harm as possible to industry."[72]

The 1998 Ordinance on the Emissions of Pollutants from Technological Processes and Technical Operations follows the requirements of the 1988 EU Large Combustion Plant Directive, but transposes the directive so as to leave some room for flexibility in achieving reductions in the emissions of air pollutants. For example, unlike the EU directive, the Polish regulation does not introduce requirements for the application of best available technology or specific desulfurization rates for large combustion units. Moreover, while the Polish regulation sets limits for small sources (with capacity below 50 MW), which is not a requirement of the EU directive, the emission limits for dust from large sources (with capacity greater than 50 MW) are considerably more liberal than the requirements of the Large Combustion Plant Directive. The Polish regulation anticipates the application of the European emission limits for dust only to installations with construction permits obtained after the ordinance comes into force.[73]

Another important difference between the Polish regulation and the Large Combustion Plant Directive relates to the definition of new plants. While the EU directive applies to all new sources, defined as those that have obtained construction permits since 1987, the cutoff date for defining new plants in the Polish regulation is March 28, 1990.[74] The Polish designation of "new plants" reflects a compromise with the power industry, which bargained for the even later cutoff date of 1996. The emission standards for new sources, which follow the requirements of the Large Combustion Plant Directive, are significantly stricter than the standards for existing sources, and are expected to impose financial difficulties on plants built between 1987 and 1996. Thus, the date for defining new sources has important economic consequences for a number of Polish power plants.[75] The difference in the definition of new and existing combustion plants was subject of negotiations for the closing of the chapter on environment in the process of EU accession.[76]

In 2001, after the adoption of the new Act on Environmental Protection, the 1998 ordinance on air emissions was amended to achieve even closer alignment with EU provisions. While the 2001 ordinance still defines new and old combustion sources differently from the EU Large Combustion Plant Directive, it mandates that all sources (old and new) which have obtained a construction permit after 1987 have to comply with standards compatible with those of the 1988 EU directive after January 2003.

Poland also adopted a more flexible approach to compliance with the Second Sulfur Protocol. The national emission reduction objectives set by the protocol are reflected in the 1996 agreement with the power industry for SO_2 reductions. The agreement assumes that the sector will reduce its emissions to 700 tons per year by 2010, which is half of the national emission ceiling for the country. By 2000, the industry had already complied with its 2010 target set in the 1996 agreement with the government, and Poland has complied with its 2010 national emission reduction target set by the Second Sulfur Protocol, despite the fact that it never did ratify the protocol. Thus as a result of its strong environmental investment program throughout the 1990s, the compliance with the national emission ceiling for Poland appeared much more feasible than anticipated in the early 1990s.

However, Poland has opted out of the costliest requirements of the Second Sulfur Protocol by regarding the plant-based emission standards specified in Annex 5 as recommendations only, and not as obligations. Annex 5 of the protocol mandates specific emission limits, technology requirements and desulfurization rates similar to the ones included in the 1988 EU Large Combustion Plant Directive to be applied to new as well as existing power plants after 2004. Compliance with such source-specific standards is costly for Polish utilities because it reduces the flexibility with which the sector as a whole can pursue improved environmental performance. That is why Poland has taken advantage of the clause stipulating that countries would apply the standards included in Annex 5 to exiting sources "as far as possible without entailing excessive costs" to interpret Annex 5 as a recommendation. This national strategy of partial and flexible adjustment to the objectives of the Second Sulfur Protocol again represents a compromise between international

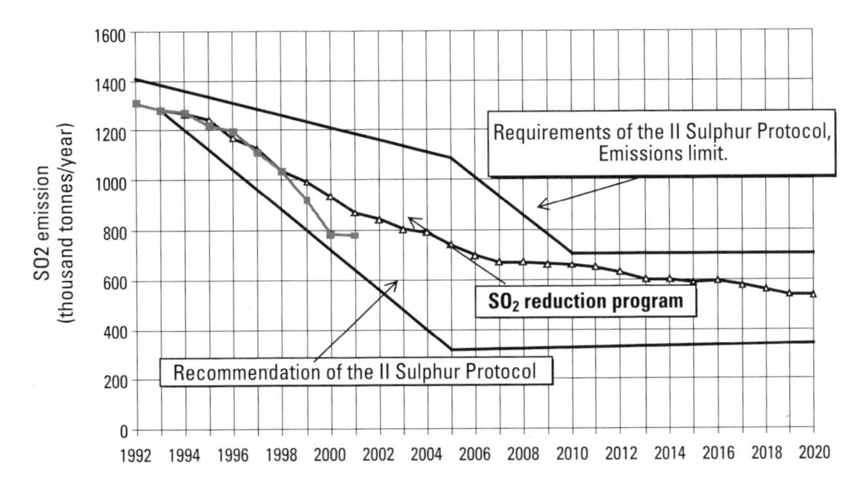

Figure 4.2
Program for SO_2 reduction in the power sector of Poland. Source: PPGC 2002.

commitments, domestic objectives, and the interests of industry. Figure 4.2 presents this point clearly by comparing the 1999 SO_2 reduction strategy of the electricity sector, the emission ceiling for the sector implied by the Second Sulfur Protocol under the assumption that the sector contributes 50 percent of all emissions in the country, and the requirements or "recommendations" of Annex 5 of the protocol. As can be seen from the graph, the anticipated emission reductions are very close to and even exceed the reductions required to achieve the national ceiling set by the Second Sulfur Protocol, but are considerably smaller than implied by the source-specific standards in Annex 5 of the protocol.

Poland thus has sought a more flexible approach to the application of European standards pertaining to acidification and transboundary air pollution, emphasizing the need to consider cost-minimization and capacity for compliance by domestic industries. The corporatist type of bargaining and agreement with industrial actors, however, succeeded in reshaping the position of industry in the course of the 1990s and stimulated a great degree of environmental investments in the sector. As a consequence, Poland achieved gradual but consistent reductions in the levels of air pollution throughout the 1990s, ultimately complying with the national emission reductions provisions of the Second Sulfur Protocol. Emissions of SO_2 were cut by 53 percent from 1990 to 2000, the

Table 4.5
Emissions of SO_2, NOx, and dust in Poland (1990–2000, thousand tons/year). Source: GUS 1997–2001.

	1990	1991	1992	1993	1994	1995	1996	1997	1998	1999	2000	Reduction 1990–2000	Reduction 1992–2000
SO_2	3,210	2,995	2,820	2,725	2,605	2,376	2,368	2,181	1,897	1,719	1,511	53%	46%
NOx	1,280	1,205	1,130	1,120	1,105	1,120	1,154	1,114	991	951	838	35%	26%
Dust	1,950	1,680	1,580	1,495	1,395	1,308	1,250	1,130	871	815	464	76%	71%

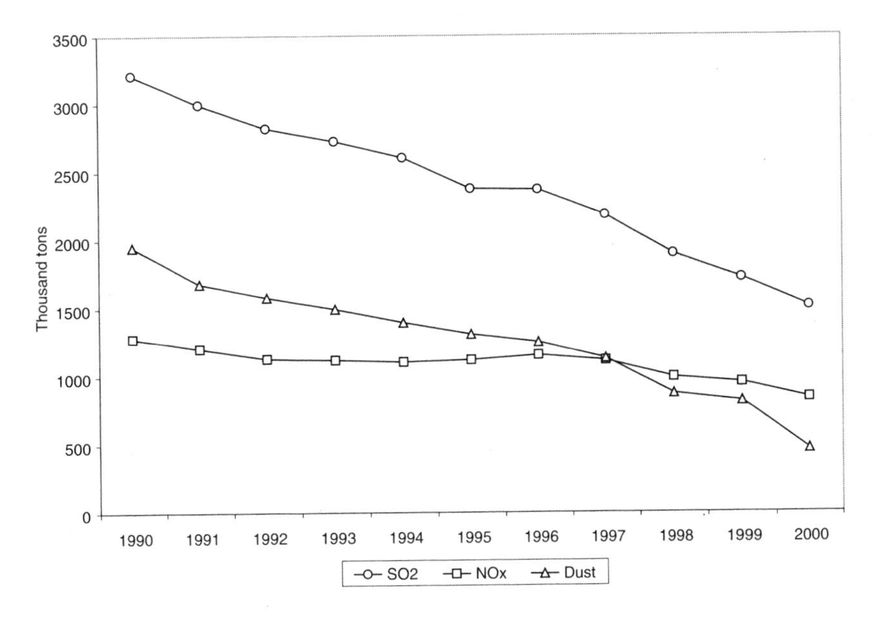

Figure 4.3
Emissions of SO_2, NO_x, and dust in Poland, 1990–2000 (thousand tons/year). Sources: GUS 1997–2001.

emissions of NOx by 35 percent, and the emissions of dust by 76 percent (table 4.5, figure 4.3).

The downward trend in acidifying emissions persisted in the years of high GDP growth after 1992 as a result of the early lock in of institutional capacity and economic incentives for environmental improvements. The driving idea behind these reforms has been an economic, flexible approach to environmental management and close, consensual decision making in which stakeholders have an important say.

Conclusion

After the collapse of communism, Poland undertook reforms in the field of air protection as part of a comprehensive environmental program driven by high public concern and by the ambition to improve its image internationally. Despite the strong commitment of the country to European integration and improved environmental performance, the influence of international norms on the air protection system was partial and gradual. In this regulatory area, unlike in the case of chemical safety, the process of regional integration presented few economic incentives for domestic actors to support the approximation of international regulations. In the Polish institutional context, the power industry enjoyed an influential position in politics and a direct veto power over the content of air pollution regulations. As a consequence, the reforms of air protection standards proceeded slowly and took close consideration of the economic interests of the sector.

Similarly to the case of Bulgaria and the Czech Republic, Polish ecological groups played a limited role in shaping regulations for air pollution control, despite the greater organizational capacity, experience and diversity of the movement compared to its Czech and Bulgarian counterparts. The most significant role of the movement again, as in the case of the Czech Republic, was limited to the early transition period when the movement influenced the ambitious reform of environmental institutions and the strong anti-nuclear sentiment prohibited development of nuclear power in Poland. The limited role of Polish environmentalists in influencing air pollution standards is attributed partly to the political demobilization of the movement during the mid 1990s. Environmental activism against air pollution in Poland was also targeted largely at the local level, against specific industrial polluters, and in relation to emerging societal problems such as increasing volume of vehicle traffic and waste incineration.

The iterated negotiation between the government and industry under the pressure of EU accession requirements, however, facilitated a bargain between broad societal concerns and particularistic interests that moved the reform of air protection forward. This bargain was underpinned by the ability of the government to support the industry's investment effort through the system of environmental funds, by setting long-term power purchase contracts, and by channeling international assistance into pollution-abatement projects. As in the case of the Czech Republic, the capacity of the government to offer financial incentives for improved environmental performance played an important role in alleviating the opposition of industrial actors to regulatory change, facilitated the evolution of their environmental strategy, and enhanced the level of compliance with international commitments. The cases of air pollution reforms in the Czech Republic and Poland thus confirm the importance of institutional factors such as the structure of veto points in the policy-making process, the interests of veto actors, and the capacity to strike compensatory bargains in determining the course of national adjustment to international pressures that create high costs for domestic constituencies.

5

Bulgaria: Harmonization without Implementation

Protests against intolerable air pollution in the city of Ruse became the symbol of anti-communist dissent in Bulgaria in the late 1980s. After the democratic changes of 1989, the abatement of air pollution, particularly in "hot spots" with excessive ambient concentrations, was identified as one of the most urgent environmental-policy priorities. Deterioration of air quality in industrial hot spots had a direct impact on the health of the population. The new international position of the country and the objective of closer integration with Western Europe required improved environmental performance. Bulgaria had one of the highest ratios of SO_2 emissions per GDP and per capita in Europe, a consequence of its highly energy intensive path of industrial development after the Second World War.

During most of the 1990s, however, Bulgaria took little action compared to the Czech Republic and Poland to address the problem of air pollution. The first post-communist governments did not capitalize on the early support of environmental causes to lock in legislative or institutional reforms with a long lasting effect. The subsequent fall in public concern and the strong veto position of the electricity sector delayed air protection reforms until the end of the 1990s. The accelerated cooperation with the EU in the second half of the 1990s facilitated the formal approximation of EU air protection standards, but the prospects for their actual implementation remained uncertain. Most of the gains in air pollution reduction in Bulgaria during the 1990s resulted from economic downturn and restructuring, and very few from the type of pollution-abatement efforts undertaken by the Czech Republic and

Poland. Thus, while the preceding two chapters illuminate institutional conditions that facilitate the adoption of international standards that impose high costs on domestic constituencies, the case of Bulgaria uncovers important institutional obstacles to the internalization of costly international rules.

Political Context: Government Instability and Weak Institutions

As in other Central and East European countries, environmental groups found a prominent place on the political agenda of Bulgaria immediately after the democratic changes of 1989. The environmental movement had been in the core of the opposition to communism. In 1987, prominent intellectuals and dissidents organized the Committee for the Ecological Salvation of Ruse in response to a widespread concern about the inaction of the communist government against the ecological crisis in the town. Ruse had been subject to massive air pollution—some from local sources, but predominantly from a Romanian factory located across the border. "The gasification of Ruse," as protesters referred to the frequent inflow of chlorine gas, caused deterioration in the health status of the population, especially children. The ecological crisis of Ruse gave rise to spontaneous rallies, growing opposition to the communist authorities, and became the first basis of a coordinated anti-communist opposition.

Bulgarian dissidents were weakly organized during the communist regime, unlike their counterparts in the Czech Republic and Poland, where prominent underground organizations such as Charter 77 in the Czech Republic and Solidarity in Poland existed since the 1970s. In Bulgaria, the opposition to the authoritarian practices of the government was limited to selected individuals, intellectuals, and academic circles, which were often targets of purges and repression. The establishment of the Committee for the Ecological Salvation of Ruse quickly became the most visible public initiative coordinating the activities of dissidents and prominent intellectuals.

The core members of the Committee for Ruse proceeded to establish the Independent Union Ekoglasnost in 1989. The platform of Ekoglasnost addressed primarily ecological objectives, but it also stated the need for democratic and economic reforms as conditions for a cleaner environment. As one study of the Bulgarian ecological movement underscores: "... although the frames used in public discourse were pure ecological ones, the unvoiced but obvious long-term aim of most participants [in Ekoglasnost], was, right from the beginning, to be a real opposition to the Communist power."[1] A rally staged by Ekoglasnost in October 1989, at the time when the international Conference for Security and Cooperation in Europe was taking place in Sofia, provoked a crackdown by police, attracted attention domestically and internationally, and precipitated the fall of the communist government.

The end of the communist regime was marked in November 1989 when a reformist faction within the Bulgarian Communist Party removed from power its long-term leader and head of state, Todor Zhivkov. The announcement of the internal party takeover was followed by a wave of mass demonstrations demanding a pluralistic and open society. The communist government pledged non-interference with independent societal activities. Soon after that, the government started round table talks with the Union of Democratic Forces, the newly formed umbrella organization representing the opposition.[2] Ekoglasnost was one of the founding groups in the Union of Democratic Forces, despite the ambivalence of many environmentalists about involvement in mainstream politics. Disagreement on the political role of the movement precipitated the creation of the Green Party in 1990 by members with explicit political ambitions. However, both Ekoglasnost and the Green Party participated in the first democratic elections under the umbrella of the Union of Democratic Forces as a way to maintain the influence of the movement in the critical period of political transition and the establishment of democratic institutions.[3]

The Bulgarian Socialist Party, the successor of the communist party, won the first post-communist elections in Bulgaria with 47 percent of the vote. The Union of Democratic Forces was second, with 36 percent of the vote. As part of the Union of Democratic Forces electoral umbrella, Ekoglasnost won 25 seats and the Green Party won 14 seats from a total of 400 seats in the parliament, together accounting for almost 10 percent of the vote in the Grand National Assembly.[4] The relatively strong presence of environmental groups in the first post-communist parliament reflected both public concern about the environment and the popularity of Ekoglasnost political figures.

The initial public and political enthusiasm for environmental change, however, was short-lived. Not long after the 1990 elections, government instability and divisions among environmentalists sharply diminished the appeal of environmental issues. The Independent Union Ekoglasnost split into two new organizations: the Political Club Ekoglasnost and the Independent Movement Ekoglasnost. Members of the Green Party also divided along ideological lines, splitting in two with the establishment of the Conservative Ecological Party.

The four successor organizations of Ekoglasnost had a distinct agenda and choice of political allies, and were soon polarized between the two dominant political forces: the Bulgarian Socialist Party and the Union of Democratic Forces. The Independent Movement Ekoglasnost remained closely associated with the Union of Democratic Forces, but steered away from direct participation in electoral politics after 1990. The Political Club Ekoglasnost had an explicitly political agenda and electoral ambitions, and with time gravitated toward a coalition with the Bulgarian Socialist Party. After 1990, many new ecological groups were established both locally and nationally, focusing predominantly on issues that generated international sponsorship such as biological diversity and conservation, public education, anti-nuclear activities, and in some cases local environmental problems.[5] The fragmentation, political maneuvering, and ideological squabbles among the most prominent environmental organizations weakened the movement and alienated the electorate from green politics.[6]

A 1995 sociological survey showed that more than two thirds of the respondents considered the environment and their health to be under strain as a result of water pollution, air pollution, nuclear power, and the accumulation of waste. At the same time only 34 percent expressed willingness to accept a lower standard of living in order to protect the environment, and nearly 80 percent felt that in the current situation "it is just too difficult for someone like me to do much about the environment."[7] Environmental objectives were overshadowed by the task of economic transformation and difficult economic reforms.[8]

The fragmentation and polarization of the political spectrum in the first half of the 1990s affected not only the environmental movement, but also political parties and institutions in Bulgaria. Compared to countries such as Poland and the Czech Republic that achieved faster economic transformation and recovery, Bulgaria did not enjoy even a short period of political consensus at the beginning of the transition in which to undertake crucial reforms. The first post-communist government, headed by the Bulgarian Socialist Party stayed in office only 11 months. The government of the Union of Democratic Forces, which won the 1991 elections, initiated reforms with the support of the IMF, but lost power in a no-confidence vote in 1992.[9] The period of political instability, frequent change of governments, and stalled economic and structural reforms continued through 1997. As a consequence, both economic and social indicators deteriorated as Bulgaria lagged in the transformation to a market economy (table 5.1). The election of a majority government of the Union of Democratic Forces in 1997 with a strong commitment to economic restructuring and reintegration in Europe accelerated economic reforms.

Between 1997 and 2000, for the first time since the democratic transition, the Bulgarian political context was characterized by relative stability. Despite the persistence of many economic and social problems during that period, the government achieved macroeconomic stability, positive economic growth, and closer cooperation with international financial institutions. In 1999, Bulgaria was invited to start negotiations for EU membership, a development that was of great significance for the government and had a strong imprint on the direction of its policy reforms.

The prolonged governmental instability during the early transition period had consequences for all policy areas, including environmental regulations. The environmental enthusiasm of the early transition period contributed to some positive changes in the making of environmental policy, but did not establish strong institutional capacity and significant legislative changes. The most significant environmental achievement of the early 1990s was the preparation and approval of the 1991 Environmental Protection Act, which provided a modern framework for environmental management. The Environmental Protection Act introduced

Table 5.1
Economic indicators for Bulgaria. Source: World Bank 2002.

	1990	1991	1992	1993	1994	1995	1996	1997	1998	1999	2000
GDP growth (%)	−9.1	−8.4	−7.3	−1.5	1.8	2.9	−10.1	−7.0	3.5	2.4	5.8
GDP per capita (1995 US$)	1,716	1,587	1,487	1,477	1,511	1,560	1,409	1,317	1,372	1,414	1,503
Inflation (%)	24	338	91	73	96	62	122	1,058	19	3	10
Unemployment (%)	1.7	11.1	15.3	21.4	20.2	16.5	14.2	13.9	12.2	14.1	16.3

the "polluter pays" principle, a requirement for Environmental Impact Assessment of economic investments, and provisions for the collection and dissemination of information on the state of the environment.[10] In 1991–92, the government also worked on a national environmental strategy in cooperation with the World Bank. The strategy was updated in 1994 and outlined national priorities and environmental-policy objectives of the type included in the Czech Rainbow Program and in the National Environmental Program of Poland.[11] Despite the clear definition of environmental problems and priority areas, however, Bulgaria made little progress in the adoption of new media-specific laws and regulations throughout most of the 1990s.

In 1995, the government of the Bulgarian Socialist Party even reversed some of the achievements of the early environmental reforms by amending of the Environmental Impact Assessment provision in the Environmental Protection Act. The amendment was made under pressure to complete a water transfer project for Sofia, for which no Environmental Impact Assessment was completed. The amended Environmental Protection Act vested the Ministry of the Environment with emergency power to approve projects "of vital interest for the population" without completing an Environmental Impact Assessment procedure and without any public participation process.[12] From the perspective of environmental advocates, the 1995 amendment of the Environmental Protection Act marked the lowest point in post-communist environmental policy making, provoking sharp protest and campaign activities. The actions of ecological groups were marginalized, however, and enjoyed little influence over the decisions of governmental agencies.[13]

The environmental conditions for EU accession played a considerable role in reinvigorating the environmental-policy reforms in Bulgaria, particularly in the late 1990s when EU membership was identified as a paramount objective for the government of the Union of Democratic Forces elected in 1997. The linkage between EU accession and environmental reforms gave environmentalists some reason for optimism at that time: "The objective of environmental protection was largely ignored by governments between 1991 and 1996. The environment is not among the priorities of the new government, either, but EU membership is. The Prime Minister will press on any issue necessary to qualify for EU accession. Indirectly, the environment will have to enter the plans of the administration."[14]

The Union of Democratic Forces maintained close political cooperation with the Independent Movement Ekoglasnost, one of the best-organized and influential environmental organizations in Bulgaria. This facilitated the interaction between the Ministry of the Environment, environmental groups, and the parliamentary majority and enhanced the influence of environmental groups on environmental policy after 1997.[15] Soon after the inauguration of the government, the parliament repealed the controversial 1995 amendment of the Environmental Protection Act that allowed investment projects to be undertaken without an Environmental Impact Assessment in cases when the "vital interests of the population are at stake." Restoring the scope of the Environmental Impact Assessment provision in Bulgarian legislation was commended as a crucial achievement by environmentalists, and ensured a high degree of harmonization with the EU directive on environmental impact assessments.[16]

The growing emphasis on preparation for EU accession negotiations also increased the leverage of the Ministry of the Environment in its efforts to promote environmental objectives, including air protection. However, the adoption and implementation of EU air pollution norms remained a highly contested political process. The application of international air emission standards, in particular, was an area of regulation that implied significant costs for Bulgaria and for its electricity industry. The interests of this strategic economic sector had a strong influence on air pollution reforms in Bulgaria, on the adoption of international standards, and the degree of their implementation.

Environmental Interests and Strategies of the Electricity Industry

Electricity production is the largest source of air pollution in Bulgaria. Thermal power and heating plants account for more than 80 percent of the SO_2 emissions, approximately 45 percent of the dust emissions, and almost 30 percent of the NOx emissions in the country.[17] As in other Central and Eastern European states, Bulgarian power plants had almost no pollution-abatement technologies installed before the democratic changes of the late 1980s. The prospect of adoption and enforcement of new air protection standards, compatible with international norms, required high levels of investment.

Electricity production is a sector of significant economic importance in Bulgaria, accounting for close to 2 percent of employment and for 16 percent of GDP. Until 2000, the industry was characterized by high vertical integration within the National Electricity Company (NEK). The NEK managed the Kozloduy Nuclear Power Plant, the thermal power stations, the majority of the hydropower stations, the regional transmission companies, and the national electricity grid. Approximately 47 percent of the electricity produced in Bulgaria is based on nuclear power, 44 percent is based on coal, and 9 percent on hydropower. About 60 percent of the primary energy resources of the country are imported, mainly from Russia in the form of nuclear fuel, crude oil, and natural gas. Most of the thermal plants rely on local brown coal, and only a small percentage of electricity (about 6 percent) is produced from higher quality imported coal. NEK has been responsible for the export and import of electricity, which represent a relatively small share of total production and consumption. Between 1992 and 1998, exports of electricity represented 1.6–10 percent of annual production, and imports of electricity accounted for 1.5–8.5 percent of annual consumption, with electricity exports increasing during the end of the 1990s due growing demand in neighboring countries (most notably Turkey and the countries of former Yugoslavia).[18]

The Bulgarian electricity sector was subject to few, if any, structural changes through most of the 1990s. The 1992 Energy Strategy Review for Bulgaria, prepared by the World Bank, recommended the demonopolization of the energy sector and improvements in its environmental performance.[19] Successive governments accomplished little along these objectives and the electricity industry remained highly integrated and heavily subsidized through most of the 1990s. This changed somewhat after the election of the 1997 Union of Democratic Forces government, which relied strongly on the support of international financial institutions that in turn required structural changes in the energy sector as a condition for continued assistance. The plan for the restructuring of the electricity industry was coordinated with the IMF as a preliminary condition for the negotiation in 1998 of a 3-year loan agreement to support economic reforms.[20]

The government adopted a new Strategy for Development of the Energy Sector and for Energy Efficiency in 1998 and the Law on Energy and Energy Efficiency in 1999. These policy and legislative changes provided the basis for the reorganization of the electricity sector, which included the separation of production, transmission, and distribution entities, and the gradual commercialization and privatization of production units. By 2000, the regional distribution companies and the majority of the thermal power plants were separated from NEK. The Kozloduy nuclear power plant was also established as an independent producer, owned by the state. NEK remained responsible for the high voltage grid and for the transmission of electricity. The 1998 Energy Strategy of Bulgaria also included gradual liberalization of the domestic energy market to prepare the sector for EU integration and to achieve compliance with the 1996 EU electricity directive. The 1998 Energy Strategy also anticipated an increase in the share of energy generated by domestic coal to 51.7 percent of production and a decrease in the share of nuclear energy to 29.7 percent of total generation as a result of the anticipated closures of some of the oldest reactors of the Kozloduy nuclear plant as requested by the EU.[21] An Energy Strategy for Bulgaria adopted in 2002 by the government of the National Movement Simeon II anticipates more aggressive liberalization of the electricity market and privatization of the majority of generation utilities.[22]

Owing to the reliance of a substantial part of electricity production on domestic brown coal and the low energy efficiency of the economy, the electricity sector is a large contributor to national emissions of SO_2 and other pollutants. The projected growth in the share of domestic coal in electricity generation would further increase the impact of the sector on the environment and the cost of air emission abatement. Brown coal, which is the only significant energy resource available domestically, has a high content of sulfur and the reduction of SO_2 emissions requires the introduction of expensive desulfurization equipment. Such equipment would be necessary for all plants fueled by domestic coal. The most significant investments would need to be made in the Maritza Iztok complex, which consists of several power generation units located in close proximity to the Maritza coal basin. The complex is considered to be one of the largest sources of air pollution in Europe, causing serious environmental hazards in its vicinity.[23]

For the Bulgarian electricity sector as a whole, high levels of investment will be necessary to achieve emission reductions compatible with EU and LRTAP standards. Industry estimates of the cost of SO_2 reduction according to the requirements of the Second Sulfur Protocol are approximately US$2.4 billion until 2010.[24] Official government reports also indicate that air protection is likely to be one of the costliest areas for compliance with EU requirements. An approximation study of the Ministry of the Environment, for example, estimates that the total cost of achieving EU air protection norms is approximately 3.02 billion euros, of which the cost of reduction of SO_2, NOx, and dust in the power sector accounts for about 2.3 billion euros.[25] Another report of the Ministry of the Environment similarly estimates the total cost of adoption of the air pollution Acquis at 4.4 billion euros, and the cost of emission reductions from stationary sources at 2.2 billion euros, of which 1.6 billion would be absorbed by the power generation sector.[26]

Thus, as in the case of Poland and the Czech Republic, the adoption of European standards for air protection imply high costs for the power generation sector in Bulgaria, while EU integration offers few immediate benefits for the sector. Moreover, the process of EU cooperation is associated with dual environmental pressure on the electricity industry both to reduce its sulfur emissions, as well as to close four of its Soviet-designed nuclear reactors. This pressure has been openly resented by representatives of the sector, who qualify EU demands as "contradictory" and "costly," requiring both reduction in the share of nuclear electricity which implies increased use of coal and further costly investments for air pollution abatement.[27] On the basis of its economic interests, the electricity sector in Bulgaria formed a strong political preference against compliance with international air pollution regulations. A statement by a representative of NEK summarized succinctly the position of the industry:

> There will be a significant increase of production costs as a result of the application of EU standards in the area of environmental protection. We cannot set aside this kind of money for investment in emission reduction technologies, since

there are more urgent tasks of restructuring and modernization. Of course, on the basis of this economic rationale the sector opposes the introduction of these standards.[28]

The electricity industry in Bulgaria has been in a position to follow closely its economic preference against the adoption and implementation of strict air emission standards. In the executive branch the sector is represented by the Committee of Energy (renamed State Agency for Energy and Energy Resources after 1999 and Ministry of Energy in 2002), which works closely with NEK and presents strongly the position of the energy lobby in the Council of Ministers. During the prolonged period of political instability and economic stagnation in Bulgaria, the financial concerns of this sector were of high importance to the government. As a consequence, the Committee of Energy and NEK were able to block any substantial legislative change in the field of air pollution during most of the 1990s and did not undertake any measures toward reducing SO_2 and NOx emissions associated with electricity production.

The Ministry of the Environment attempted several times to "tie its hands" behind international commitments to advance the air protection agenda. But such commitments by themselves had little effect on the environmental strategies of the electricity industry, in the absence of any real positive or negative incentives to apply abatement measures. The pollution fees and fines were low during the 1990s. There was no system of permits to control air emissions, and the effort to bring chronic violators into compliance was weak.[29] Given the strong political position of the sector, the electricity industry in Bulgaria could not be forced, as in the Czech Republic, to consider options for improved environmental performance. Nor could the Bulgarian government provide, as in the cases of both Poland and the Czech Republic, financial or other compensatory mechanisms to mitigate the cost of environmental expenditures. The capacity of the National Environmental Fund remained low (table 5.2) and other mechanisms for preferential environmental financing were weakly developed. Through most of the 1990s, the electricity industry in Bulgaria largely ignored the issue of air pollution from power plants and resisted any attempt on the part of the Ministry of the Environment to strengthen air emission standards and their implementation.

Table 5.2
Expenditures of the National Environmental Protection Fund of Bulgaria (1995–2000, million US$). Data in million BGL provided by REC 2001, converted into US$ by author using average annual exchange rates provided by the Bulgarian National Bank.

	1995	1996	1997	1998	1999	2000
Total	6.3	8.7	4.1	25.4	26.4	25.9
Air pollution	1.6	1.1	0.4	2.5	1.5	2.2
Percent air pollution	25	13	9	10	6	8

The structural reforms in the electricity sector and the strong governmental emphasis on preparation for EU accession negotiations after 1997 put environmental issues more squarely on the agenda of NEK and the Committee for Energy. The approximation of EU environmental standards was no longer just a priority for the Ministry of the Environment, but also became a priority for the government and its leadership. In addition, the sector lost some of its monopolistic influence in the process of structural and ownership changes. With the separation of the power plants from NEK and preparations for their long run commercialization and privatization, NEK was no longer directly responsible for environmental improvements and became more willing to agree on long-term strategies for environmental planning.

The National Energy Strategy of 1998 included a special section on environmental policy reflecting the need to take into account the international obligations of the country. The Committee of Energy also worked out in 1999 an "Action Plan for Meeting the Commitments of the Republic of Bulgaria under International Environmental Agreements, based on the National Energy Strategy." The Action Plan identified the restructuring of the sector, improved energy efficiency, and the construction of desulfurization equipment as the main elements of a program for the reduction of polluting emissions. The document included a list of projects for plant-rehabilitation and desulfurization that would be necessary to achieve international objectives. However, the Action Plan contained no clear designation of implementation responsibility or a strategy for financing, and projected emissions reductions based on anticipation of restructuring and investment to be undertaken by future owners and operators of the plants.[30]

The formal adoption of environmental programs in the electricity sector at the end of the 1990s was not coupled with significant investments in pollution reduction measures. By 2000, none of the thermal power stations met the requirements of modern technology and pollution control. The only sulfur-abatement project undertaken by the sector during the 1990s was the construction of desulfurization equipment in two units of the Maritza Iztok II complex. The project was financed with assistance from the European Bank for Reconstruction and Development (EBRD) and the European Investment Bank, and did not become operational until 2002. As a consequence of the low level of pollution-abatement investment in the power sector, the relative contribution of the power sector (electricity and heat) to total air pollution remained high, accounting for more than 80 percent of the emissions of SO_2, 32 percent of the emissions of NOx, and 34 percent of the dust emissions in the country in 1999. The trend of emission reductions has been uneven and driven to a great degree by economic recession in the early 1990s and restructuring in the second half of the decade (table 5.3, figure 5.1).

The Bulgarian electricity industry managed to consistently reduce only its dust emissions. This was a result of the use of electro-filters in most of the thermal electricity generation units.[31] The decline in the industry's emissions of SO_2 that resulted from the sharp economic recession at the beginning of the transition was reversed after 1992. The SO_2 emissions from the power generation sector increased between 1992 and 1994, after which emissions remained high despite a trend of slow decline since 1995 (figure 5.1). A more significant decline in the total emissions of the power sector between 1998 and 1999 is noted chiefly as a result of emission reductions in district heating utilities after the substitution of heavy fuel oil with fuels with lower sulfur content. The emission of NOx from the power industry also increased after the initial drop achieved by 1993, before starting to slowly decline again toward 1993 levels. The lack of a consistent environmental program in the electricity sector has had profound implications for air quality management in Bulgaria during the

Table 5.3
Emissions of SO_2, NOx, and dust from the power industry (electricity and district heating) in Bulgaria, 1990–2000 (thousand tons/year). Source: National Center for Environment and Sustainable Development 1990 through 2001 and Council of Ministers of the Republic of Bulgaria 1999 and 2000.

	1990	1991	1992	1993	1994	1995	1996	1997	1998	1999	2000	Reduction 1990–2000	Reduction 1992–2000
SO_2	1,216	1,102	920	1,119	1,311	1,243	1,196	1,143	1,032	789	794	35%	14%
NOx	84	77	89	66	83	71	67	65	66	63	51	39%	43%
Dust	190	172	181				149	118	95	49			73%

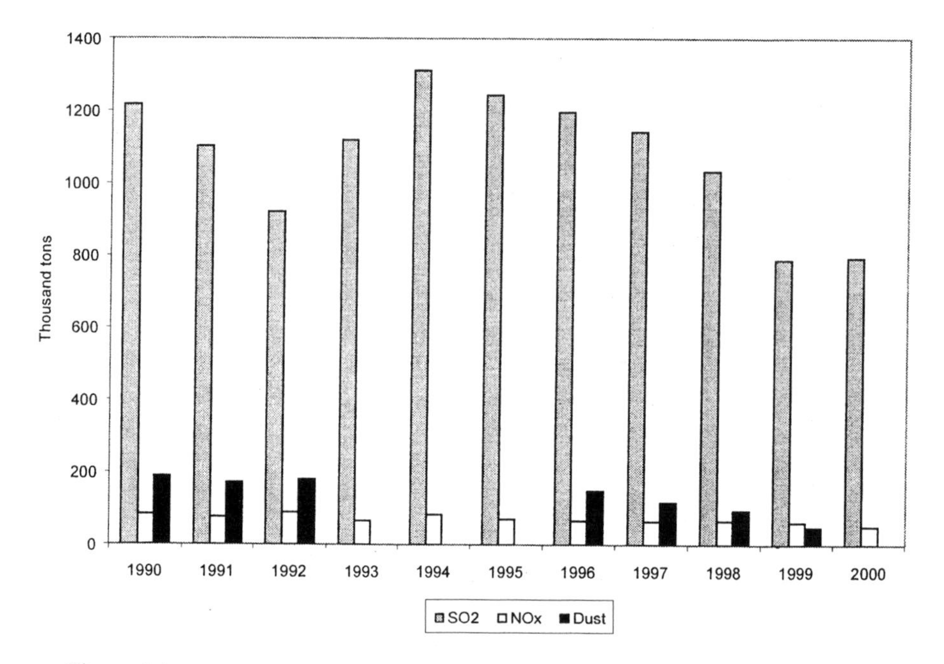

Figure 5.1
Emissions of SO_2, NOx, and dust by the Bulgarian power industry (electricity and district heating), 1990–2000 (thousand tons/year). Sources: National Center for Environment and Sustainable Development 1990–2001; Council of Ministers of Republic of Bulgaria 1999, 2000.

1990s and for national efforts to achieve closer compliance with European standards.

Air Pollution Policy

The high levels of air pollution in post-communist Bulgaria resulted from more than 45 years of intense and inefficient industrialization. In the period 1939–1989, the share of industry of the Bulgarian GDP grew by 15–57 percent, with heavy industry accounting for 65 percent of industrial production.[32] The industrialization of the country, which is relatively poor in mineral resources, took place primarily with the support of subsidies both from the Bulgarian government and from the former Soviet Union. As a consequence of the fast and inefficient growth of the Bulgarian economy under communism, pollution levels increased sharply, especially during the 1970s and the 1980s. In 1990, the country emitted 2,020 tons of SO_2.[33] The most serious deterioration of air quality was observed around big power plants and industrial facilities, forming pollution hot spots, where allowable concentration levels were regularly exceeded. Industrial hot spots were often located in close proximity to big cities such as Sofia, Ruse, Plovdiv, Bourgas and Dimitrovgrad. According to estimates of the National Center for Environment and Sustainable Development, in 1990, about a third of the Bulgarian population lived in "hot-spot" areas, characterized by highly polluted air (table 5.4).

The public health implications of air pollution in Bulgaria are well documented.[34] In towns with high concentration of ambient pollutants, the risk of respiratory diseases and health defects among children is significantly greater than in other areas of the country. In the vicinity of the Maritza Iztok power plants, for example, where three large coal-burning power stations are concentrated and air pollution is particularly heavy, the percentage of respiratory diseases is about 15 percent higher than the national average.[35] Correlation between acute respiratory disease and air pollution has been documented in other industrial cities as well, including Ruse, Devnya, Kremikovtsi, Asenovgrad, Sofia, and Dimitrovgrad.[36] Epidemiological studies have also found consistent evidence of higher incidence of infant mortality, birth defects, pulmonary disease, and growth retardation among children living in regions where the acceptable standards for dust, sulfur dioxide, and lead have been exceeded for years.[37]

Because of its damaging health effects, air pollution was identified early in the transition period as one of the top three priorities of Bulgarian environmental policy, together with water pollution and soil contamination.[38] The public concern with the fate of the citizens of Ruse and of other industrial hot spots also raised expectations that the problem would be addressed as a priority by the democratic government. Air pollution, moreover, was a visible transboundary problem for Bulgaria, which was a source and a recipient of significant flows of acidifying pollutants. In response to domestic and international concerns, the environmental administration developed early in the transition period a national strategy for air protection. This strategy envisaged the adoption of new legislation and executive regulations, strengthening the system of monitoring and control, extending the role of economic instruments, and enhancing the capacity of regional inspectorates and municipalities to enforce air pollution standards.[39]

Despite the priority of air protection in the programs of the Ministry of the Environment, the progress of reforming the system of air pollution management was slow and uneven. The political instability and economic crisis of the early 1990s undermined the urgency of cleaning up the environment and strengthened the political position of industrial interests over those of the environmental ministry and public advocates. In dealing with the reduction of air pollutants, the Ministry of the Environment faced strong opposition from the powerful energy lobby, which had a de facto veto power in the government and ability to block any substantial regulatory changes. As a consequence, the government did not undertake significant legislative or institutional reform in the air protection sector until 1996.

In 1991, the Ministry of the Environment adopted an ordinance setting emission limits for different industrial branches, including power plants, but without effective instruments for enforcement.[40] There was no permit system, which would allow for a regular control of plant-based emissions, nor the installation of continuous monitoring equipment at any of the major pollution sources. The penalties for exceeding emission limits were low, tended to lag behind inflation, and did not provide an incentive for improved performance. According to an OECD report, the total air pollution fines collected in 1993, for example, was approximately 700,000 euros, which is a relatively low amount given the increased levels of emissions in Bulgaria after 1992.[41] Only a few industrial enterprises, subject to direct public protest and threat of lawsuits, adopted pollution-abatement measures. Such examples include a lead and zinc smelter near Plovdiv and a fertilizer plant near Devnya, where visible local polluters attracted continuous media coverage, public action, and the scrutiny of regional inspectorates.

The 1964 Act on Air Protection held little force after the democratic changes and was subject to reformulation. The Ministry of the Environment prepared a new Law on Air Protection as early as 1992, but the approval of the draft law by the Council of Ministers and its submission to the Parliament was delayed almost indefinitely. The draft had to be coordinated with interested ministries and institutions, whereby all suggestions made in writing had to be reflected in the version submitted for discussion at the Council of Minister. The Council of Ministers then had the authority to approve the proposed legislation and to submit it to the Parliament, or to return it to the Ministry of the Environment for further changes.[42] In this policy-making structure, the energy sector held a strong political position, being directly represented in the executive branch by the Committee of Energy and having a substantial influence over the decisions of the Council of Ministers. From the perspective of environmental advocates, the government acted as a "hostage" to the interests of the energy lobby in making decisions on environmental and nuclear safety issues related to the sector.[43]

The Ministry of the Environment used international commitments as the only mechanism that could push its environmental program forward. Bulgaria signed an association agreement with the EU in 1993 and applied for EU membership in 1995. Bulgaria was also a member of the LRTAP convention for limiting the transboundary flow of air pollutants. The signing of the Second Sulfur Protocol in 1994, which roughly coincided with the preparations for the third pan-European Conference of Environmental Ministers to be held in Sofia in 1995, was grasped by the Ministry of the Environment as an opportunity to place the draft air protection legislation more forcefully on the government's agenda. Given the importance of reductions in transboundary air pollution to Western European states, the Ministry of the Environment was able to argue that the adoption of the new Act on Clean Air by the government and by the parliament would send an important signal of commitment to European cooperation. Many participants in the policy-making process at that time described the Act on Clean Air as a "political and diplomatic act for the 1995 Environment for Europe conference in Sofia. A way to demonstrate a step in the right direction in front of the international community."[44]

The fact that the approval of the draft clean air legislation by the Council of Ministers coincided with the beginning of the Environment for Europe conference in Sofia is indicative of its symbolic significance.[45] The Act on Clean Air was introduced in the parliament within two days of its adoption by the government. In presenting the law on the parliamentary floor, the representative of the Parliamentary Committee on the Environment emphasized that the Act on Clean Air addresses a priority problem, tries to make up for years of delays in the legislative program concerning the environment, and demonstrates a serious intention to honor the international commitments of Bulgaria.[46]

The version of the Act on Clean Air that was approved by the government and presented to the parliament did not introduce any immediate threat of costly regulation for industries. The Act on Clean Air is a framework law that outlines the principles of air protection and the functions of control institutions. It provides only the legislative basis for the introduction of more specific air quality and emission standards by executive regulations. Moreover, article 10 of the draft clean air legislation included a provision for granting temporary emission limits to large sources, which "due to the type of fuel used and the level of technology cannot meet the legal standards."[47] Temporary emission limits could be approved by the Ministry of the Environment, or in the case of the energy sector by the Council of Ministers. They could extend up to 5 years, and are determined on the basis of the technical condition of the enterprise and negotiation between interested parties.

The provision for temporary emissions limits was hotly debated in the parliament. Representatives of the majority Bulgarian Socialist Party justified the temporary emissions clause by emphasizing that when "we have to choose between the desirable and the possible, it is better to be realistic."[48] Representatives of the Ministry of the Environment also made an argument that without the introduction of temporary limits, there is a risk that no limits will be met and the legislation will become a "paper law" without implementation.[49] The opposition Union of Democratic Forces recognized that the timely adoption of an Act on Clean Air would improve human health and Bulgaria's international image, but rejected the inclusion of temporary emission limits. Parliamentary representatives of the Union of Democratic Forces argued that article 10 de facto absolved enterprises from any responsibility to control pollution and undermined the effectiveness of the whole air protection legislation. The opposition also pointed out that the provision for temporary emissions reflected the interests of the energy sector and other big polluters and undermined considerations for human health and the right of the public and local authorities to influence the quality of their living environment. The opponents of the temporary limits provision maintained that article 10 of the Act on Clean Air introduced an element of arbitrariness in the regulatory system and violated the "polluter pays" principle embedded in the Environmental Protection Act. This provision allowed for temporary increase in pollution levels without any penalty, without specifying ex-ante how temporary emission limits would be determined, and without tying such decisions to enforceable programs for plant-level emission reductions.[50]

There were proposals on the parliamentary floor to remove article 10, as well as a more moderate proposition for amendment of article 10 to include a clause that tied temporary emission limits to compliance with emission reduction programs, backed by implementation and enforcement instruments.[51] These proposals were discussed in the Parliamentary Committee of the Environment with the participation of the Ministry of the Environment, the Ministry of Construction and Regional Development, and NEK. None of the proposals related to the provision on temporary emissions were approved by the environment committee or by the parliamentary majority during the second reading of the law. The Act on Clean Air was passed in May 1996.[52] This new legislation advanced the objective of reforms in the air protection sector, but also reflected the influence of strong industrial interests by leaving space for enterprises to opt out, at least temporarily, from national emission standards.

The implementation of the Act on Clean Air required the adoption of governmental ordinances establishing admissible air quality and emission limits. The Ministry of the Environment commissioned several projects intended to bring about a closer agreement with the energy sector on new air protection standards. In 1995–96, the Ministry of the Environment and the Committee of Energy signed a Protocol of Cooperation and initiated a project to consider options for establishing new air emission standards in accordance to the requirements of the Second Sulfur Protocol. However, the project did not result in any substantive agreement, regulatory changes, or investment activities targeted at emission reductions. Another study, commissioned by the Ministry of the Environment and sponsored by the World Bank, presented a least cost analysis of compliance with the Second Sulfur Protocol and had no influence on the strategy of the Committee of Energy and NEK.[53] The World Bank assessment recommended coal-to-gas conversion in the sector as the cheapest strategy to achieve SO_2 reduction targets. Since such a strategy implied increased dependence of Bulgaria on imports of Russian natural gas, it was rejected by the energy sector as irrelevant and lacking understanding of the local energy conditions.[54] In the absence of any financial stimulus for environmental investments in the sector or any other compensatory measures to offset the expected high cost of air emission reductions, the electricity industry remained unwilling to support environmental reforms or to consider measures for improved performance.

The slow progress of Bulgaria in its environmental reform and in managing air pollution was reflected in the 1997 Opinion of the European Commission on the application of the country for EU accession: "Bulgaria's environmental problems are very serious, and have not been effectively addressed. Bulgaria has high emissions of air pollutants, in particular sulfur dioxide and particulates originating from thermal power plants, heavy industry, domestic heating and motor vehicles. Local air quality poses significant risks to human health. . . . Air emission standards are inadequate for some substances and existing standards are not uniformly enforced."[55] In 1997, Bulgaria's sulfur emissions per capita were 168 kg, compared to 80 kg per capita for Hungary and 71 kg per capita for Poland.[56] A 1998 report of the government estimated that as much as 30 percent of the population continued to be affected by high levels of air pollution in 1995, approximately the same share as in 1990 (table 5.4).

The acceleration of the EU harmonization agenda after 1997 affected the regulatory activity in the field of air protection. Harmonizing EU environmental legislation became the dominant objective of the Ministry of the Environment and facilitated its relations with other ministries. The establishment of 30 working groups dealing with EU approximation, including group 22 (on the environment), enhanced inter-ministerial coordination. Because of the importance of air pollution management to EU members, the Ministry of the Environment was able to press on with the schedule for adoption of air pollution regulations

Table 5.4
Populations of "hot spot" cities in Bulgaria. Source: National Report of the Republic of Bulgaria on the Reduction of SO_2 and Dust Emissions 1998.

	1990	1991	1992	1993	1994	1995
Population (millions)	2.77	2.76	2.62	2.63	2.62	2.61
Percentage of total population	31.97	32.09	30.89	31.06	31.16	31.19

implementing the 1996 Act on Clean Air in ways that followed the broad requirements of EU directives.[57] In this work, the Ministry of the Environment was assisted by a number of EU-funded projects targeted specifically at the approximation and implementation of the air pollution Acquis. Some of these projects resulted in the formulation or amendment of several air pollution regulations, including the 1998 EU directive on emissions from large combustion sources.[58]

While between 1991 and 1997 the Ministry of the Environment issued only one regulation in the field of air protection and adopted the Act on Clean Air but without implementing regulations, between 1997 and 2000, the ministry drafted and adopted 12 new ordinances on air pollution pursuant to the 1996 legislation. The Act on Clean Air was also amended in 2000 to achieve closer correspondence with the EU Framework Directive on Air Quality. The air pollution regulations issued by the Ministry of the Environment in cooperation with other relevant ministries followed the requirements of EU legislation and covered different aspects of air pollution management, including norms for ambient concentrations of polluting substances, air quality management and monitoring, measuring and recording emission levels, emissions of VOCs, emissions from large stationary sources, limits on the content of lead, sulfur and other substances in fuels, and temporary emission limits.[59]

Regulations 2 and 15, issued in 1998 and in 1999 by the Ministry of the Environment, the Ministry of Health, the Ministry of Industry, and the Ministry of Regional Development, deal explicitly with emissions from large stationary sources. Regulation 15 sets emission limits for SO_2, NOx, and dust from new large combustion plants, following the standards of the 1988 EU Large Combustion Plant Directive. An important difference between the EU directive and the Bulgarian ordinance on air emissions from large combustion sources is the definition of new plants. According to Regulation 15, new plants are those that were built or for which permits for major reconstruction were obtained after January 2000. The EU directive defines new installations as those that obtained a construction permit after 1987. Regulation 2 refers to emission limits from stationary sources more broadly, including large combustion plants which were built or obtained an Environmental Impact Assessment permit for major reconstruction before 2000. This regulation also sets technology and air emission standards compatible with the standards of the 1988 Large Combustion Plant Directive and of the Second Sulfur Protocol. However, the regulation also allows existing sources to apply for exemptions from the mandated standards on the basis of meeting temporary emission limits, whereby existing sources are defined as those that were built or acquired Environmental Impact Assessment permits for reconstruction before 1998.[60]

Thus, while regulations 2 and 15 set standards for air emissions from stationary sources compatible with European requirements, these ordinances also provide a possibility for all sources that existed at the time of their adoption to avoid, at least in the short and the medium term, the application of national standards. Enterprises built before 1998 can apply for temporary emission limits. Only sources that began operation or undertook major reconstruction after 1998 are under strict obligation to comply with technology and emission standards.

The 2000 amendments to the Act on Clean Air retained the provision for temporary emission limits, but introduced some changes that set a requirement for the adoption of emission reduction programs in conjunction with the temporary exemptions from the national standards. The procedures for governmental decisions on temporary emission limits are specified in ordinances 1 and 3 of the Ministry of the Environment, issued in 1998. Ordinance 3 applies exclusively to enterprises in the power generation sector, allowing existing power plants that "cannot meet emission limits due to their resource base or level of technology" to apply for temporary standards negotiated between interested parties and approved by the Council of Ministers. While working under temporary emission limits, such enterprises are not required to pay penalties or taxes for higher levels of pollution, and are subject to all pollution fees only if they do not meet the temporary norms.[61] The regulation does not set a time limit for operating under temporary emission agreements, although such decisions usually extend from 5 to 8 years.

The temporary emission limits specified by the Bulgarian air protection legislation provide the energy sector and other polluting enterprises with a cushion against the requirements of strict air pollution regulations, without which the formal adoption of EU and LRTAP norms would probably have been impossible. The Ministry of the Environment avoided the veto of the energy sector by applying strict performance requirements primarily to future operations. It also used the temporary emissions limits provision of its regulations to increase the pressure on utilities to adopt and eventually comply with a program for emission reductions. From the point of view of the environmental administration, this is a considerable improvement compared to the early 1990s, when the ministry and its enforcement institutions had virtually no recourse to press for emission abatement in the power sector.

As the electricity industry began internal restructuring in 2000, NEK and the Committee of Energy were no longer so concerned with the cost of regulations that were to be applied in the long run. The emission reduction plans projected by the Committee of Energy in its 1998 Energy Sector Strategy reflected projections for modernization and reconstruction made by future owners or operators of electricity utilities, whose identity was still largely uncertain.[62] Restructuring in the electricity sector and the temporary emission limits provision, thus, made the cost of air pollution regulation appear more diffuse and uncertain and muted the opposition of a strategic industrial player. The government was able to achieve a relatively high level of formal approximation of international standards in the air protection sector by deferring to the future the difficult questions of implementation.

Compliance with International Standards

After years of inaction, Bulgaria adopted almost wholesale the principles of EU air protection regulations at the end of the 1990s. In 2000, a review of air protection presented by the Ministry of Environment Bulletin concluded that with the amendment of the Act on Clean Air and the adoption of 12 implementing ordinances, the approximation of the EU air protection Acquis would be largely completed by the end of 2000, with the exception of some of the latest EU directives that would be adopted by 2002.[63] The formal approximation of EU and LRTAP air emission and technology standards, however, was achieved at the price of leaving a wide loophole for all existing enterprises to temporary opt out of these standards.

The total reduction of air pollution emissions followed an uneven path during the 1990s, being largely an artifact of the economic downturn in the first half of the decade and restructuring later on. In the case of SO_2, for example, there was a significant reduction in emissions between 1990 and 1992, which coincided with the sharp recession of the economy. In the wake of the slow economic recovery after 1992, this tendency was reversed, and there was a new increase in SO_2 emissions. SO_2 emissions began to decline slowly again after 1995; a more significant decrease between 1998 and 1999 was due in part to the diminished use of heavy fuel oil in district heating plants and in part to a slightly different method of emissions inventorization. The total reduction in SO_2 emissions between 1990 and 2000 was 51 percent, reaching 982 tons a year, which is a more significant reduction than the national emission target for 2010 set by the Second Sulfur Protocol (table 5.5). However, compared to 1992 sulfur emission levels, Bulgaria has achieved only a 10 percent reduction by 2000, indicating that a large part of the reductions are a result of economic slowdown and restructuring during the 1990s (table 5.5, figure 5.2). Dust emissions decreased 56 percent between 1992 and 2000, and NOx emissions 19 percent.

The limited investment in pollution abatement by the power sector puts in doubt the ability of Bulgaria to achieve the source-based emission and technology standards mandated by the 1988 Large Combustion Plant Directive and the Second Sulfur Protocol, let alone stricter standards associated with the 2001 Large Combustion Plant Directive. If the economy maintains a path of moderate growth after 2000, it will be increasingly difficult for the country to maintain reductions in total air pollution emissions without investment in energy efficiency and

Table 5.5
Emissions of SO_2, NOx, and dust in Bulgaria, 1990–2000 (thousand tons/year). Source: National Center for Environment and Sustainable Development 1990 through 2001 and Council of Ministers 1999 and 2000.

	1990	1991	1992	1993	1994	1995	1996	1997	1998	1999	2000	Reduction 1990–2000	Reduction 1992–2000
SO_2	2,020	1,678	1,093	1,422	1,482	1,497	1,420	1,364	1,251	942	982	51%	10%
NOx			229	238	326	266	259	224	223	202	185		19%
Dust			423	382	353	358	306	265	233	185			56%[a]

a. reduction 1992–1999.

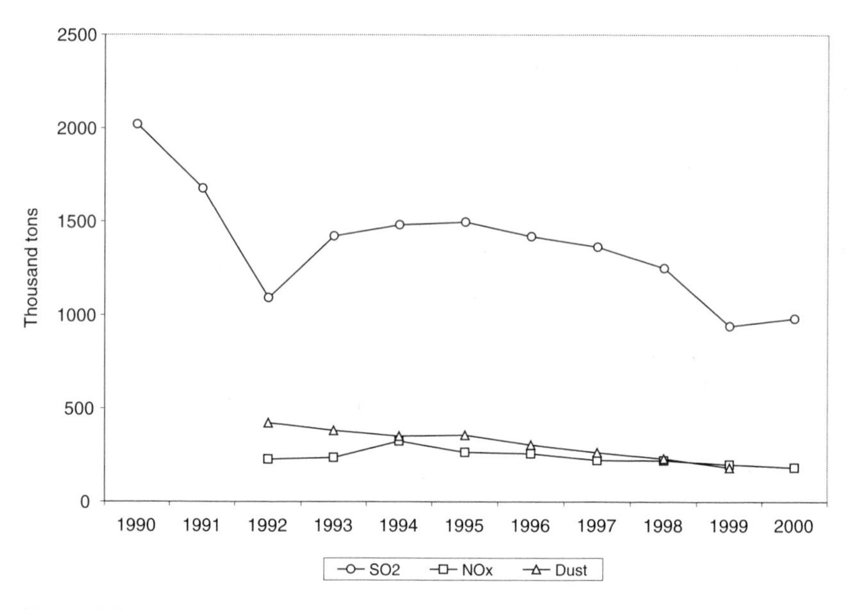

Figure 5.2
Emissions of SO_2, NOx, and dust in Bulgaria, 1990–2000, (thousand tons/year). Sources: National Center for Environment and Sustainable Development 1990–2001; Council Ministers 1999, 2000.

the types of pollution abatement undertaken by the Czech Republic and Poland during the 1990s. The need for pollution abatement will increase further as the share of coal-based electricity generation increases after the shutdown of reactors 1 and 2 of the Kozloduy nuclear plant in 2002, and the projected shutdown of its reactors 3 and 4 in 2006 or close after that date.

Conclusion

In the aftermath of communism, Bulgaria, like the Czech Republic and Poland, inherited a significant air pollution problem and was hard pressed by international institutions to reduce its emissions and achieve standards compatible with those mandated by EU regulations and the LRTAP protocols. During the first decade of post-communist transition, however, government instability and powerful industrial interests delayed regulatory and institutional reforms. The government commitment to EU integration and to industrial restructuring facilitated the formal adoption of EU and LRTAP air pollution regulations at the end of the 1990s. The high level of formal harmonization was not matched, however, by any realistic projections for implementation of stricter standards in the near and medium term future. Enterprises could opt out of the regulations on the basis of temporary emissions agreements. National emission reductions have been inconsistent and uncertain, and high ambient concentration of pollutants continues to cause health problems in industrial hot spots.

The course of air protection reform in Bulgaria and its adjustment to international standards was influenced considerably by the position and strategies of the electricity industry, as in the cases of Poland and the Czech Republic. The energy sector in Bulgaria, like its counterpart in Poland, had a strong veto power over the adoption and the contents of air pollution regulations. While both the Czech and the Polish governments were able to provide compensation for the cost of regulation through policy bargains and investment assistance, Bulgaria's capacity for compensation of the cost of environmental improvements was minimal, as was the enforcement capacity of environmental institutions. Bulgaria, therefore, could not offer either positive or negative incentives for the electricity sector to change its environmental performance. The use of international institutions as a commitment mechanism facilitated some aspects of environmental reform, but was not sufficient to change the preferences or strategies of industrial actors, which anticipated high costs of compliance with international air protection standards.

The weaker institutional capacity for compensation and enforcement in Bulgaria reflected a broader trend of slow and incomplete post-communist reforms. The same fundamental factors that delayed economic restructuring and stabilization also hampered the "lock in" of environmental institutions and policy changes early in the transition period. The resulting delays in economic restructuring and prolonged economic hardship further diminished the capacity and political will for environmental improvements. In sum, Bulgaria lacked the domestic institutional framework to facilitate compliance with costly international commitments.

The influence of the environmental movement was also marginalized shortly after the initial high public concern with the environment. The role of environmental groups in shaping air protection regulations was relatively limited in Bulgaria. While isolated protest activities and local pressure influenced the behavior of a few industrial enterprises, there was no strong, coordinated presence on the part environmental groups in the making of air protection regulations. As in Poland and the Czech Republic, Bulgarian environmentalists had limited information about and input in EU harmonization processes. The agenda of green groups was increasingly driven by international donor assistance and priorities emphasizing predominantly biodiversity, access to information, anti-nuclear activities, and monitoring of international development banks.

Despite the low domestic capacity for environmental management and the concentrated opposition to international air pollution standards in Bulgaria, the pull of EU accession did have some influence on the reform of air protection policies. European commitments entered domestic politics to affect the strategic interaction between governmental institutions and societal interests. In the three cases of air pollution reforms considered in this book, EU institutions and the process of integration influenced national policies by providing a commitment mechanism for reform-oriented governments. This effect was more limited and dependent on domestic circumstances compared to the influence of integration on chemical safety policies, where international institutions and markets created strong economic incentives for adoption of international standards and transformed profoundly domestic regulatory politics. Thus, the comparison of air protection and chemical safety reforms in Eastern Europe illuminates different mechanisms and different extents of international influence. The air pollution cases, furthermore, demonstrate that in instances where international agreements require costly adjustments, the scope of their effect is strongly dependent on the characteristics of domestic institutions, and on the presence of facilitating factors that mitigate the resistance of affected interests and enhance the leverage of the supporters of reforms. Recognizing the different mechanisms of international influence on domestic politics is critical for addressing issues of the effectiveness of international cooperation, and for managing the effects of regional and global integration.

Conclusion

The chapters of this book examine ten years of environmental policy reforms in three post-communist countries in Central and Eastern Europe. The case studies presented detailed policy insight of the making of chemical safety and air pollution regulations in Poland, the Czech Republic, and Bulgaria during a period of economic and political transition and active integration in EU markets and institutions. The analysis demonstrates that it is increasingly difficult to understand domestic policy choices without taking into account international pressures. It specifies the differential effects of integration on domestic environmental interests and the role of domestic institutions to account for the degree of national adjustment to EU environmental norms. Beyond illuminating the forces that shape the environmental future of Central and Eastern European states, the analytical framework and findings of this study have implications for several broader research agendas in international relations and environmental policy: the internationalization of domestic politics, the influence of international institutions and transnational networks, and the impact of global integration on environmental and other regulations.

Internationalization of Domestic Politics

This book contributes to the political science scholarship that since the 1970s has made a concerted effort to bridge the divide in the study of international and domestic politics. The political economy literature, in particular, has made great strides in uncovering the impact of international trade, finance, and economic shocks on national economic and social policies.[1] Recently, EU studies also turned their attention to the "Europeanization" of domestic politics in member states under the impact of EU institutions.[2] This book extends the open-economy line of analysis to environmental politics to examine how EU markets and institutions affect environmental protection in transition states.

Understanding the interplay between domestic and international factors in shaping environmental policy is particularly important. Environmental regulations often address complex problems that affect multiple layers of actors and governance. It is therefore surprising that so far relatively few studies have made a conscious effort to specify the mechanisms of interaction between international and domestic environmental political incentives.[3] This study offers a method for analyzing the internationalization of environmental politics in the context of integration. It combines institutionalist and interest-based perspectives to internationalization in order to specify mechanisms and conditions for differential external effects. The study reveals a truly transnational picture of environmental policy making in Central and Eastern European states—a picture in which international incentives, institutions, and actors interact closely with domestic political processes.

The internationalization of domestic environmental politics is most clearly demonstrated in the chemical safety cases. In this arena, international markets, institutions, and transnational associations precipitated a significant change in domestic interests, coalitions, and policies. As the theoretical framework anticipated, the linkage between environmental regulations and exports to EU markets provided strong commercial incentives for the export-oriented chemical industry to adjust its environmental interests and to support the reform of chemical safety regulations. The direct pressure and involvement of transnational organizations further reinforced the change in industry-government relations. Their influence facilitated the rapid adoption of international chemical safety standards in Bulgaria, the Czech Republic, and Poland, despite differences in domestic institutions and capacity among the three states.

The internationalization of chemical safety regulations in Central and Eastern Europe was evidenced directly and indirectly by numerous policy details, including such extraordinary facts as the availability of draft chemical legislation in the English language, the direct involvement of

EU institutions and international business organizations in the discussion of legislative proposals, and in the case of Bulgaria, the preparation of the first draft of the Law on Chemicals by the chemical industry association in cooperation with foreign experts.

The cases of air pollution reforms did not present conditions for such a profound influence of international markets and institutions on domestic politics. The electricity generation industry, the principal target of regulations that limit acidifying emissions into the air, was not involved extensively in international trade and the linkage between EU markets and environmental norms did not provide a motivation for the sector to support the adoption of strict international standards which implied high costs without offsetting benefits. Regional integration thus did not result in a change in the environmental interests of the electricity industry.

Even in the field of air protection, however, the influence of integration was visible, and stronger than might have been predicted on the basis of narrow consideration of the economic costs and benefits associated with it. The linkage between broader foreign policy objectives and environmental norms within EU institutions provided a commitment mechanism for the environmental administrations of Central and Eastern European governments to advance air protection reforms. In the absence of strong commercial incentives for harmonization, however, the effectiveness of this institutional mechanism of EU influence varied across the three states I examined, depending on the degree to which domestic structures helped override industrial veto or facilitated a compensatory bargain.

The specification of the market and non-market mechanisms of international influence on domestic regulations has profound implications for the comparative study of Central and Eastern European politics beyond the environment. It can be used in further studies to illuminate how EU and other international pressures influence domestic politics in a variety of areas, including social protection, food safety, telecommunication regulations, company law, judiciary reforms, and other elements of the post-communist reform agenda. After the collapse of the communist regimes, the scholarly community interested in this part of the world responded with a rich comparative literature on the dynamics and outcomes of post-communist transitions.[4] This literature engages issues of democratic consolidation, institutional choice, party system formation, economic reforms, as well as environmental, social, health, and other policies. The analytic focus of most post-communist studies has been placed primarily on domestic political actors, national histories, and institutions. The theoretical approach and findings of this book clearly indicate that it is time to look beyond the boundaries of the state for sources of influence on political interests and choices in Central and Eastern Europe.[5]

The theoretical approach used in this book can be used to analyze the influence of international markets and institutions in other emerging markets. The cases from Central and Eastern Europe show that the lower level of institutionalization of emerging economies, their rapid integration in international markets and institutions, and their relatively high dependence on international markets and resources is likely to magnify the significance of external influences. Understanding the mechanisms of international influence, its differential effect across economic actors and regulatory areas, and the significance of domestic institutions is therefore critical for the successful management of international pressures in emerging markets. This approach makes it possible to identify who loses and who wins from integration and the adoption of international standards, which in turn provides cues to understanding the political hurdles and facilitating factors shaping national policies.

One central proposition advanced by this study, for example, is that international markets underpinned by a set of rules create positive incentives for actors that benefit from integration to support the diffusion and domestic harmonization of international standards. Positive incentives for the adoption of higher international standards include easier access to international markets as a result of removing non-tariff barriers to trade, reduced transaction costs, reduced societal pressure, diffusion of supporting technology, and advantage over domestic competitors. If this argument is true beyond the cases examined in this book, we should be able to observe a trend of diffusion of environmental, safety, and even social rules from large regulated markets to emerging markets. Moreover such transnational diffusion of rules should be driven to a large extent by transnational economic interests and organizations rather than solely by state actors and agreements. Some evidence of such trends is already provided by accounts of the exportation of chemical safety norms to developing countries by multinational corporations,[6] the diffusion of regulatory standards across North America and the EU,[7] and attempts by global corporations to institutionalize and promote a set of voluntary social and environmental standards.[8] Further comparative, cross-regional research can illuminate the relative significance of transnational market and normative pressure for domestic governance (particularly in weakly institutionalized regimes), the limits of such influence, and the implications for states and domestic regulations.

International Institutions and Transnational Networks

Another central topic of international relations research engaged by this book is how international institutions and norms affect state policies. Two sets of theoretical approaches have emerged in analyzing institutional influence in international relations. The rationalist, state-centered perspective emphasizes the role of institutions in reducing the transaction cost of cooperation, helping overcome collective action problems, and providing mechanisms for credible commitments and issue linkage, that in turn can alter the payoffs for states and their willingness and ability to undertake a particular course of action.[9] Scholarship in this tradition also increasingly takes into account the role of domestic politics to explain variation in institutional effects.[10]

A second powerful perspective in the institutionalist literature, social constructivism, views international rules, norms, and ideas not so much as mechanisms to constrain state behavior given a set of predetermined interests, but as forces fundamentally constitutive of identities, behavior, and even interests in international politics.[11] An important part of this literature examines the role of societal and policy networks for the diffusion and influence of norms and ideas across the domestic and international realms.[12] The literature on international environmental cooperation exemplifies both the rationalist, state-centric approach, as well constructivist and society-centric approaches in seeking empirical evidence of the effects of global environmental regimes.[13] For most part, however, the state-centered and society-centered approaches in regime studies developed in parallel. Only recently have scholars attempted to bring the two perspectives together in a more active dialogue.[14]

This study of EU influence on environmental policies in Central and Eastern Europe demonstrates that both inter-governmental mechanisms of institutional commitment and transnational mechanisms of norm diffusion were at play. The linkage between geo-strategic, economic, and environmental objectives within the context of EU institutions provided a rationale for accession states to commit to the adoption of even the costliest EU standards. The high credibility of this commitment, in turn, influenced domestic bargaining and provided a mechanism for environmental administrations to press for reforms. But even when government commitment is strong and credible, as in the case of the Eastern European commitment to EU accession, the case studies revealed that taking account of domestic politics is critical for understanding differences in the influence of international rules. Moreover, the analysis revealed that domestic preferences and institutions are not always exogenous to international regimes. Understanding the mechanisms through which regimes affect domestic actors and capacity increases the analytical leverage in disentangling institutional effects.

The analysis also shows that the exclusive focus on EU institutions and governmental commitments characteristic of most studies of EU enlargement captures only half of the story of EU influence on domestic environmental policies. EU integration and rules affect environmental politics in accession states also through channels associated with transnational markets and networks. The cases of the chemical industry highlight the importance of international markets and business networks in promoting and even monitoring the adoption of EU environmental rules, particularly in regulatory areas that affect highly integrated regional markets and industries. In other areas of EU environmental regulations not subject to this research, such as environmental impact assessment, relatively well-organized transnational advocacy networks have played a role in translating EU norms into domestic rules. Moreover these and other examples indicate that when strong transnational and inter-governmental mechanisms coexist, international institutions exert the greatest influence on domestic politics and on national policies.[15]

The potential for a mutually supportive role of inter-governmental regimes and transnational networks and interests raises a number of more challenging questions for further research. Under what conditions do transnational networks supportive of the rules of inter-governmental regimes emerge? Are transnational networks an unintended consequence of international regime building or do they emerge in parallel to inter-governmental agreements driven by social and market interests? Should inter-governmental agreements try to foster transnational markets and organizations that would promote the principles they agree to? Or would such mechanisms represent a growing infringement on the basic international norm of state sovereignty?

In the cases considered here, transnational forces were at work supporting the adoption of chemical safety norms, but not the adoption of air pollution regulations. Transnational advocacy networks are relatively well organized in the areas of biodiversity protection, human rights, and access to information, and relatively absent in field of local industrial pollution, water and air protection, and soil erosion. One claim of this study, reminiscent of earlier functionalist arguments about EU integration and of theories of transnational relations,[16] is that a high degree of market integration and internationalization of production create conditions (both interests and organizational structures) for the diffusion of rules that pertain to these markets. The pooling of sovereignty and governance to supranational institutions in the EU has also been accompanied by increasing reliance on a variety of transnational actors and networks (courts and lawyers, trans-European organizations, and societal interests) for the implementation and monitoring of European rules.[17] This book indicates that there is clearly a need to accord more analytic attention to the role of markets, transnational networks, and private codes as increasingly prominent transnational forces of political influence and governance.[18] Systematically analyzing the nexus between transnational and inter-governmental mechanisms of influence in international regimes would allow the international relations scholarship to provide answers to important questions that are not only of academic but also of practical interest for effective global governance.

Global Integration and the Environment

Finally, this study contributes to the debate on the impact of growing regional and global integration on the ability of states to protect environmental and other social values.[19] The globalization and environment debate has been highly polarized. On one hand, it is asserted that the pressures of international trade and investment would lead to a regulatory race to the bottom, particularly in emerging market economies, which have weak institutions and compete for foreign investment and markets.[20] Others have countered race to the bottom arguments by pointing out that global integration can have a positive effect on environmental regulations by raising income in open economies and the demand for environmental protection, by promoting more efficient production, by facilitating clean technology transfer, and the diffusion of higher voluntary and non-voluntary standards from advanced to emerging markets.[21]

This book presents a story in which the inclusion of the formerly closed communist economies in regional and world markets was clearly paralleled by a significant strengthening of their environmental protection systems under the influence of EU institutions. But is not the European case, the reader may wonder, in many ways a unique example of high level of economic integration governed by a dense set of supranational institutions? Isn't it the exception rather than the rule? Indeed, despite the fact that the Treaty of Rome (1957) established the European Economic Community without mentioning the objective of environmental protection even in a "side agreement" akin to that of the North America Free Trade Agreement, in the course of its deepening economic and political integration the European Community and later the EU adopted a complex set of supranational environmental rules. The EU represents the strongest case of delegation of regulatory powers to a supranational institution, which is unlikely to be replicated in the near future in other regions or globally. And it was in fact the linkage between the economic benefits of EU integration and the adoption of EU environmental rules that provided one of the most important stimuli for environmental reforms in accession countries. Is it possible then to draw broader implications from this example about the relationship between international markets and environmental protection in emerging economies?

While the theory of transnational influence on domestic environmental politics was developed here in the specific context of European integration, the focus on the market and institutional mechanisms of international influence has broader relevance for understanding the nexus between integration and environmental regulations. One broadly generalizable argument advanced here is that the effect of integration on the regulatory preferences of industrial actors varies across sectors and firms. This implies that the analytical focus in the globalization and environment debate should strive to shift away from the good-bad dichotomy toward specifying the conditions for different impacts of global integration as a more solid basis for managing its consequences.

On the basis of the present study it is also possible to identify several general conditions for a positive, negative or neutral effect on the environment by international markets, and conditions that may stimulate the diffusion of higher standards from more regulated to less regulated areas. First, the importance of the linkage between market and normative pressure for changing the environmental position of the chemical industry in transition states suggests that only markets underpinned by a set of regulations or voluntary standards can provide incentives for improved environmental performance. Moreover, the active normative pressure from transnational networks (business, consumer or advocacy) is often important for amplifying positive environmental incentives that could be associated with regulated international markets. While, for example, Central and Eastern European exporters to the EU may face strong pressures to adopt EU standards both by markets and EU organizations, those exporting primarily to the markets of the former Soviet Union where few regulations and normative pressures exist have no similar incentives. Thus, it is the strong orientation of Central and Eastern European economies to the EU area that underlies the overall positive effect of openness on certain industrial sectors.

Many race-to-the-bottom arguments assume that the international market place is largely unencumbered by rules and norms. The reality is more complex than this, however. While global markets are not governed by strong supranational institutions, the largest and most attractive trading areas from the perspective of emerging economies (North America, the EU, and Japan) are relatively highly regulated by domestic and, in the case of the EU, by supranational standards. Participation in such markets creates both economic incentives and normative pressure that can stimulate change in domestic preferences and practices. In many instances, including agricultural trade, wood and paper processing, automobile emissions, and food safety, regulations of large industrialized markets have exerted pressure for the adoption of similar practices elsewhere in the world.[22] Increasingly, a set of voluntary standards with a market bite and a set of transnational organizations are also established seeking to promote voluntary codes as a way to protect consumers or level the playing field and preempt stricter government regulation. Such governance pressures are another element of global interdependence likely to facilitate the transnational diffusion of environmental norms and counterbalance pressures for a race to the bottom.

The example of the electricity industry shows, however, that not all economic agents have positive incentives to adopt higher international standards even if they are part of economies that are open to international markets densely governed by supranational and domestic norms. Transnational pressure for improved environmental performance is only effective on those economic agents (predominantly large export-oriented firms and sectors) that benefit most from integration and have the capacity to internalize the cost of regulation or transfer it to consumers. By contrast, for largely domestic sectors like electricity generation in Central and Eastern Europe during the 1990s, there are few commercial incentives to adopt costly international standards. Such incentives are similarly nonexistent or even negative for import competing firms and sectors, for which additional regulation would imply no benefits and further loss of competitiveness, and may indeed provide a strong rationale for lobbying in favor of laxer regulations. Understanding the conditions and the limits of a positive environmental effect of global integration thus implies that trade itself is unlikely to work an environmental miracle, making open economies cleaner. Governments of emerging market economies will need to make a considerable regulatory effort in all areas of environmental protection, while emphasizing different types of compliance incentives.

Policy Implications

The anticipation of differential environmental effects of integration across industrial actors and policies has important implications for designing environmental policies in the context of globalization. By recognizing the existence of positive market incentives for environmental performance, policy makers can seek to magnify the positive link between markets and environmental norms through capacity building, improved provision of information, formal adoption of international standards, and better access to credits for technology transfer and innovation. Such policies would strengthen the conditions for a beneficial influence of economic openness on the environment. International assistance that seeks to promote improved environmental management in emerging markets can also strive to move away from only government-to-government programs and to strengthen institutions that will increase the reputation and societal pressure on economic entities. The three cases of chemical safety reforms demonstrate unambiguously that international support can illuminate the market benefits of regulation, can compensate quite effectively for domestic institutional gaps, and thus can facilitate a faster and more effective course of reforms.

The air pollution cases demonstrate, on the other hand, that when integration or international regulations create losing sectors domestically, national regulators can mobilize a host of international and domestic resources to move the environmental agenda forward. The detailed examination of different strategies and the extent of national compliance with international air emission standards confirms that institutional capacity and the ability to work out compensatory bargains can be crucial in enhancing reforms and adjustment to international agreements and globalization pressures.

Economic prosperity can clearly strengthen the institutional capacity to cope with international pressures, but as the example of the Polish Funds for Environmental Protection demonstrates, innovative institutional design can lock in financial resources for environmental improvement even in conditions of difficult economic reform. The environmental funds in Poland provided targeted assistance for environmental investment and helped to move the air protection agenda toward reconciliation of international, environmental, and industrial objectives. In the Czech Republic, the government and the electricity industry worked out a completely different compensatory bargain that emphasized policy linkage and reflecting the weaker blocking power of the sector and the opportunities for policy tradeoffs afforded by the domestic context. Thus, rather than wither the state in the context of globalization, strong institutional capacity and the ability to adopt flexible policies are likely to be crucial for responding to global pressures in environmental protection and other policy areas.

As a consequence of international pressure and incentives, as well as strong efforts at domestic institution building in the early transition periods, Central and Eastern European countries achieved remarkable gains in environmental regulation during a period of economic transition. Of the examples considered in detail here, the Czech Republic, Poland and Bulgaria adopted within a little more than 5 years, and are on the course of implementing EU and OECD regulations on chemical safety that were negotiated and developed over the course of decades by advanced capitalist economies. Even more surprisingly, Poland and the Czech Republic and their electricity sectors achieved within less than 10 years a relatively high level of compliance with some of the most demanding air emission standards. Most of these gains were not solely a result of an economic downturn but of consistent institution building and targeted investment. Only Bulgaria, where environmental protection and financing institutions remained weak during the 1990s, lagged in the implementation of strict air emissions standards.

These examples counter expectations of the early 1990s that the adoption of EU regulations is likely to be a nearly impossible task for Central and Eastern European states and their formal compliance may be accompanied by limited actual implementation. Significant gains have been achieved in other environmental areas as well, including access to information, increasing water protection, biodiversity, and the elimination of lead in petrol. The relatively rapid adoption of EU standards in these relatively poor countries, compared to their West European counterparts, also challenges the myth that environmental protection is a privilege of the rich.[23] Targeted institution building, international assistance, the development of technology, and the emergence of transnational networks can significantly reduce the hurdles to less developed countries in adopting higher standards established by international treaties or developed states. The advantage of step-wise adoption of costly international standards, first by more developed areas and then by poorer states, while enhancing domestic capacity to manage externally imposed costs is an important lesson of this study for practical regime building efforts.

The cases presented here also indicate that the institutional forms most appropriate for a given domestic context will depend on national histories and the constellation of political interests and opportunities. While we can talk of characteristics that facilitate policy adjustment, we should be cautious about the notion of "optimal institutions" often advanced explicitly or implicitly by international development or environmental assistance programs. Taking account of domestic politics and structural realities in devising capacity building programs may seem cumbersome and time consuming for international institutions and negotiators, but is likely to enhance the effectiveness of cooperation. With this cautionary note and with guarded optimism about the environmental future of Central and Eastern European states, we end this journey, which traveled the intricate road of post-communist transition, regional integration, and environmental regulations.

Appendix
List of Interviews

Bulgaria

Government Institutions and Representatives

Ministry of Environment and Waters of the Republic of Bulgaria

Deputy Minister for European Integration, Sofia: March 1998

Department of International Relations, Sofia: March 1998

Department for Industrial Pollution and Waste Management, Sofia: March 1998, November 2002 (telephone interview)

Department of Air Protection, Sofia: June 1997, March 1998, July 1999, July 2000, August 2002, December 2002 (e-mail)

Director, Legal Department, Sofia: June 1997

Director for EU Integration and Economic Instruments, Sofia: June 1997, March 1998

Others

Former Deputy Minister of the Environment and Waters, Sofia: March 1998, July 2000

Former official from the Ministry of the Environment and Waters, Sofia: March 1998

National Assembly of the Republic of Bulgaria, Chairman of the Committee for Environment, Sofia: August 2000

National Center for Environment and Sustainable Development, Director, Sofia: January 1998, March 1998

Non-Governmental Organizations

Green Balkans, Sofia: July 1997

Green Balkans, Plovdiv: July 2000

Borrowed Nature, Sofia: March 1998 (telephone interview)

Za Zemjata, Sofia: July 2000

Regional Environmental Center (REC) Bulgaria, Sofia: July 1997, March 1998

Informacionen Centar po Ekologia, Sofia: August 2000

Former member of Ekoglanost, Sofia: August 2000

Demetra, Sofia: July 2000 (telephone interview)

Industry Organizations

Bulgaria Chamber of Chemical Industry (BCCI), Sofia: March 1998, August 1998, February 1999 (telephone interview), July 2000, August 2002

Clean Industry Center of the Bulgarian Industry Association, Sofia: July 1997, March 1998, August 1998, July 2000

National Electricity Company (NEK), Department of Environmental Management, Sofia: March 1998, June 2000

Committee of Energy, Department of EU Integration, Sofia: March, 1998

Committee of Energy, Department of Environment, Sofia: June 2000, August 2000

Energoproekt, Sofia: March 1998

Czech Republic

Government Institutions and Representatives

Ministry of Environment of the Czech Republic

Department of Ecological Risks and Monitoring, Prague: November 1997, February 1999 (telephone interview), March 1999 (e-mail exchange)

Department of International Relations, Prague: November 1997, January 1998

Deputy Minister, European Integration, Prague: November 1997, January 1998

Department for Air Protection, Prague: November 1997

Others

Ministry of Foreign Affairs of the Czech Republic, EU Integration Unit, Official responsible for environmental approximation, close associate of the late Minister of the Environment Josef Vavrousek, Prague: November 1997

Office of the President of the Czech Republic, foreign policy advisor and member of the environmental movement, Prague: November 1997

Non-Governmental Organizations

Greenpeace Praha, Prague: November 1997

Green Circle, Prague: November 1997, January 1998

Regional Environmental Center (REC) Czech Republic, Prague: November 1997

Rainbow Movement: November 1997 (telephone interview)

Rainbow Movement Praha, Prague: November 1997

Program for Energy Efficiency, Prague: November 1997

SEVEN, Prague: November 1997

Industry Organizations

Association of the Chemical Industry of the Czech Republic (ACICR), Prague: November 1997, January 1998

Czech Business Council for Sustainable Development (CBCSD), Director, Prague: November 1997, January 1998

Czech Environmental Management Center (CEMC), Prague: November 1997, January 1998

Former Head of the Environmental Protection Department of the Czech Electricity Company (CEZ), Prague: November 1997

Academic and Research Institutions

Professor Bedrich Moldan, Charles University, Former Minister of the Environment, Prague: November 1997

Dr. Josef Sejak, Czech Environmental Institute, Prague: November 1997

Poland

Government Institutions

Ministry of Environment, Natural Resources and Forestry of the Republic of Poland

Deputy-Director, Department of International Relations, Warsaw: February 1998, June 2000

Expert, Department of International Relations, Warsaw: June 2000

Director, Environmental Policy Department, Warsaw: February 1998

PHARE advisory office on EU law approximation, Warsaw: February 1998

Director, Air Protection Department, Warsaw: June 2000

Environmental Policy Department, experts on air protection, Warsaw: February 1998

PHARE legal adviser, Warsaw: February 1998

Head of EU Accession Team, Department of International Relations, Warsaw: February 1998

Others

National Fund for Environmental Protection and Water Management, Expert, Warsaw: February 1998

Director, Air Protection Unit, National Fund for Environmental Protection and Water Management, Warsaw: June 2000

Ministry of the Economy of the Republic of Poland, Department of Energy, Warsaw: February 1998

Council of Ministers of the Republic of Poland, Committee for European Integration, Advisor to the Chief Negotiator, Warsaw: June 2000

Head of the Chancellery of the President of Poland, Warsaw: February 1998

Chancellery of the Sejm, Office for Research of the Sejm, the Parliament of the Republic of Poland, Warsaw: February 1998, July 1999 (e-mail exchange), and June 2000

Parliament of the Republic of Poland, the Sejm, Committee of the Environment, Warsaw: February 1998

Non-Governmental Organizations

Institute for Sustainable Development, Warsaw: February 1998, June 2000

Polish Ecological Club, Warsaw: February 1998, June 2000

Regional Environmental Center (REC) Poland, Warsaw: February 1998

Social Ecology Institute, Warsaw: February 1998

EU Information Center, Warsaw: June 2000

Industry Representatives

Polish Power Grid Company (PPGC), Department of Environmental Protection, Warsaw: February 1998, June 2000, November 2002 (e-mail)

Polish Chamber of Chemical Industry (PCCI), Department of Environmental Protection, Warsaw: February 1998

Industrial Chemistry Research Institute, Warsaw: March 1998

Business Council for Sustainable Development of Poland, Chairman, Warsaw: February 1998

Academic and Research Institutions

Head of Office of Harvard Institute for International Development Poland, Economic Adviser to the Ministry of the Environment, Natural Resources and Forestry, Warsaw: February 1998

Professor Tomasz Zylicz, Department of Economics, Warsaw University, Economic adviser to the Ministry of the Environment, Natural Resources and Forestry, Warsaw: February 1998, June 2000

Dr. Boleslaw Jankowski, Energysys: Warsaw, February 1998

Nofer Institute for Occupational Medicine, Department of Chemical Risk and Safety (responsible for preparing the text and revisions of the Draft Act on Chemical Substances), Lodz: August 1999 (telephone interview), November 2002 (e-mail)

European Union

European Commission

Directorate General for Environment, country officers for Bulgaria, the Czech Republic, and Poland, Brussels: November 1997, April 1998

Directorate General for Environment, Head of Air Protection Unit, Brussels: November 1997, April 1998

Directorate General for Energy, Brussels: April 1998

Other

Manager for Central European and Regulatory Affairs, CEFIC, Brussels: April 1998.

Notes

Introduction

1. The organization's name was European Community (EC) until the ratification of the Maastricht Treaty, in 1992, which introduced the name European Union (EU). I use "European Community" when referring to events that occurred before 1992.

2. In this book, the terms "Central and Eastern European" and "Central and Eastern Europe" refer to the ten post-communist countries that applied for EU membership during the 1990s: Bulgaria, the Czech Republic, Estonia, Hungary, Latvia, Lithuania, Poland, Romania, Slovakia, and Slovenia. Cyprus and Malta also applied for EU membership in the 1990s; not being former communist states, they are not objects of the study.

3. Zielonka and Pravda 2001.

4. Bofinger 1998; Faini and Porters 1995; Henderson 1999; Mayhew 1998; Redmond and Rosenthal 1998; Schimmelfennig 2001; Sedelmeier and Wallace 2000; Van Brabant 1999.

5. Souces: Polish Agency for International Investment, accessed in October 2002 via http://www.paiz.gov.pl; Bulgarian National Bank, accessed in October 2002 via http://www.bnb.bg; Czech National Bank, accessed in October 2002 via http://www.cnb.cz.

6. European Council 1993.

7. For broad reviews of the enlargement process, see Baun 2000; Dawson and Fawn 2002; Grabbe and Hughes 1998; Henderson 1999; Mayhew 1998; Nicoll and Schoenberg 1998; Preston 1997; Schimmelfennig 2001; Sedelmeier and Wallace 2000. On East-West cooperation after the Cold War, see Hoffmann, Keohane, and Nye 1993.

8. Schimmelfennig 2001.

9. Cameroon 2002.

10. Cameron 2002; Grabbe 2001; Crisen and Carmin 2002; Holzinger and Knoepfel 2000; Jacoby 1999; Lynch 2000; Tang 2000.

11. Crisen and Carmin 2002; Holzinger and Knoepfel 2000; Lynch 2000; World Bank 1998; World Bank 2001.

12. Souce of data: Garvey 2002. A similar estimate of 100–120 billion euros is provided by the web site of the European Commission, accessed in November 2002 via http://europa.eu.int. Estimates of the costs of adoption of the EU environmental Acquis by accession candidates vary and the European Commission has provided financing for a range of studies that assess these costs by sector of regulation, as well as for the entire Acquis. For a more detailed analysis on the range of economic costing studies and their application in environmental reform in Central and Eastern Europe, see Botcheva 2001.

13. Commission of the European Communities 1997d; Meacher 1998; World Bank 1997, 1998, 1999, 2001.

14. The PHARE program provided assistance to all accession countries.

15. Figure provided by the web site of the European Commission, accessed in November 2002 via http://europa.eu.int.

16. Beckman 2002; Bell 2002; Hicks 2002; Baker and Jehlicka 1998; Holzinger and Knoepfel 2000; REC 2000.

17. For a list of all transition periods requested or granted as of 2002, see Droll 2002.

18. Lynch 2000; Caddy 1997; Jacoby 1999; Kramer 2002; Bell 2002.

19. Prominent works in these different traditions include Haas 1968; Hoffmann 1966; Moravcsik 1991, 1998; Pierson 1996; Pollack 1997; Burley and Mattli 1993; Tsebelis and Garrett 2001; Eichengreen and Frieden 1998; Wallace and Wallace 2000. For a concise review of different theoretical approaches to explaining EU integration, see George 1996. Majone 1996 and Heritier 1999 exemplify prominent works that advance the understanding of regulatory policy in the EU.

20. Heritier et al. 2001; Cowles, Caporaso, and Risse 2001; Borzel and Risse 2000; Borzel 2000; Duina 1999; Haverland 2000; Haas 1998; Haigh 1990; Glachant 2001; Grabbe 2001; Knill and Lehmkuhl 1999.

21. For overviews of the main approaches and arguments, see Heritier et al. 2001; Cowles, Cporaso, and Risse 2001.

22. Eising 2002; Knill and Lehmkuhl 1999.

23. Haas 1968.

24. For a review of this literature, see Keohane and Milner 1996.

25. Rogowski 1989; Frieden 1991; Milner 1988; Frieden and Rogowski 1996.

26. Alt 1987; Bates 1997; Garrett 1998; Garrett and Lange 1996; Goldstein 1986; Gourevitch 1986; Hall 1986; Katzenstein 1978; Kitschelt et al. 1999a; Rodrik 1999; Rogowski 1987; Berger 1996.

27. Prominent examples in this tradition include the following: Brown Weiss and Jacobson 1998; Haas, Keohane, and Levy 1993; Hurrell and Kingsbury 1992; Keohane and Levy 1996; Victor et al. 1998; Young 1989, 1998, 1999.

28. See e.g. Andersson 1999; Andrews 1993; Baumgartl 1993, 1997; Brown, Weiss, and Jacobson 1998; Cole 1998; Hicks 1996; Jendroska 1998; Klarer and Moldan 1997; Vari and Tamas 1993.

29. Choucri 1993; DeSombre 2000; Garcia-Johnson 2000; Haas 1989; Keck and Sikkink 1998; O'Neill 2000; Steinberg 2001.

30. Becker (1983), Posner (1974), and Stigler (1971), for example, advance an economic theory of regulation that identifies the interests of regulated sectors as the single most important determinant of regulatory choice. Even studies that move away from such an extreme economic reductionism to take into account the role of politics and institutions in shaping regulatory policies treat the interests of relevant socio-economic groups as central explanatory factors. See e.g. Hahn 1990; Keohane, Revesz, and Stavins 1996; Majone 1996; McCubbins, Noll, and Weingast 1989; Oye and Maxwell 1995; Yandle 1989.

31. The definitions of actors' preferences and strategies follow the conceptualization of Frieden (1999).

32. The costs stem from additional investments needed to achieve the required pollution reductions. Environmental regulations can also be associated with selective benefits such as direct subsidies, price fixing, and control over entry by rivals. See Keohane, Revesz, and Stavins 1996; Oye and Maxwell 1995; Stigler 1971.

33. Prominent works in this tradition include Rogowski 1989, Frieden 1991, and Milner 1988. Rogowski builds on the Stolper-Samuelson theorem, which anticipates that owners of factors in which a country is abundant relative to the world will win from trade while owners of scarce factors will lose, to examine how interests split along class lines (land, labor, capital) and the implication for trade policies. Frieden follows the Ricardor-Viner approach in trade theory to specify the distribution implications of external economic shocks across sectors in Latin American countries. Milner illuminates how trade integration affects firm-level interests and the trade policy implications of these political dynamics. For a broader review of interest-based approaches on the domestic political effects of trade and external economic shocks, see Frieden and Rogowski 1996.

34. Bofinger 1998; Halpern 1995; Inotai 2000; Inotai 1999; Tang 2000.

35. Smith 2000.

36. Bofinger 1998; Tang 2000.

37. Inotai 2000.

38. Ibid.; Bofinger 1998; Faini and Portes 1995; Halpern 1995; Smith 2000.

39. Interview with industry associations in Central and Eastern Europe; Jakubczyk 1997; Jha, Markandya, and Vossenaar 1999; Zylicz and Holzinger 2000. On the role of large regulated markets in "ratcheting up" the

environmental and safety standards in trading partners through the commercial interests of exporters, see Eliste and Fredriksson 1998; Vogel 1995.

40. Interview at CEMC, Prague, 1998.

41. Jha, Markandya, and Vossenar 1999.

42. Ibid.; Mol 2001; Grant 1993.

43. Others have also noted a tendency by multinational enterprises to diffuse environmental standards across borders in response the societal and shareholder pressure, and in the interest of having a uniform regulatory environment, the ability to preempt further regulations, and to export their cleaner technology. See Grant 1993; Garcia-Johnson 2000; Mol 2001; OECD 1999b.

44. See Haas 1968.

45. On the role of business organizations in EU environmental policy making, see Cowles 2001; Grant 1993; Grant, Matthews, and Newell 2000; Mol 2001.

46. For examples, see Mol 2001; De Simone and Popoff 1997; Garcia-Johnson 2000; Ruggie 2002.

47. CEFIC 1996b.

48. OECD 1998a,b; REC 1998a; CEFIC 1998.

49. OECD 1998a; REC 1998a.

50. Evans, Jacobson, and Putnam 1993; Martin 1992; Moravcsik 1998.

51. Grabbe 2001; Carius 2002; Jacoby 1999; Lynch 2000; Holzinger and Knoepfel 2000; Andersson 1999; Baumgartl 1997.

52. On the Europeanization literature see Heritier et al. 2001; Cowles, Caporaso, and Risse 2001; Borzel and Risse 2000; Borzel 2000; Duina 1999; Haverland 2000; Haas 1998; Haigh 1990; Glachant 2001; Grabbe 2001. For insitutionalist analyses of open-economy politics, see Alt 1987; Bates 1997; Garrett 1998; Garrett and Lange 1996; Goldstein 1986; Gourevitch 1986; Hall 1986; Katzenstein 1978; Rogowski 1987; Berger 1996.

53. Haverland 2000; Cowles; Caporaso, and Risse 2001; Heritier et al. 2001; Tsebelis 1995, 2002.

54. Alt 1987; Hall 1986; Cameron 1978; Katzenstein 1978; Gourevitch 1986; Rodrik 1999; Garrett 1998.

55. Bates 1997; Goldstein 1986; Hansen 1990; Rogowski 1987.

56. Tsebelis (1995, 2002) examines most thoroughly the role of veto points in the policy process. On the significance of veto points for the adoption of EU policies by EU member states, see Haverland 2000; Cowles, Caporaso, and Risse 2001; Heritier et al. 2001.

57. Alt and Gilligan 1994; McCubbins, Noll, and Weingast 1989; Rogowski 1987.

58. Alt 1987; Cameron 1978; Garrett 1998; Katzenstein 1985; Kitschelt et al. 1999a; Rodrik 1997, 1999.

59. Katzenstein 1985.

60. Garrett 1998.

61. Gourevitch 1986; Katzenstein 1985.

62. Heritier 1999; Keohane, Revesz, and Stavins 1996.

63. Haas, Keohane, and Levy 1993; Keohane and Levy 1996.

64. REC 2001.

65. *Warsaw Voice* 1999.

66. Gourowitz 1999; Haas, Keohane, and Levy 1993; Finnemore 1996; Keck and Sikkink 1998; Martin and Sikkink 1993; Young 1998.

67. Beckman 2002; interviews with environmental organizations in Central and Eastern Europe (see appendix).

68. On the influence of donor organizations on the agenda and the organizational methods of environmental movements during the post-communist transition, see Baker and Jehlicka 1998; Jancar-Webster 1998; Waller 1998; Hicks 2002.

69. Hicks 2002; Jehlicka and Tickle 2002; Bell 2002; interviews with environmental groups in Central and Eastern Europe (see appendix).

70. King, Keohane, and Verba 1994.

71. For country profiles, see the UN Statistics Division's web site: http://unstats.un.org.

Chapter 1

1. CEFIC 1996a, 1997a; Garcia-Johnson 2000; ICCA 1996.

2. Year 2001 data from ACICR 2002.

3. Commission of European Communities 2000.

4. ACICR 1998a.

5. Year 2001 data from ACICR 2002.

6. Year 2001 data from ACICR 2002.

7. ACICR 1997a, 1998a.

8. ACICR 1998a.

9. CzechInvest 1997.

10. Commission of the European Communities 2000.

11. ACICR 2002.

12. The ACICR was established in 1990 as a branch of the Industry and Transport Union. It is a voluntary association of 81 chemical companies, which account for more than 80% of the production of the Czech Republic's chemical sector. See ACICR 1994, 1996.

13. ACICR 1996, 1997a; interviews at ACICR, November 1997.

14. ACICR 1998a; interviews at ACICR, November 1997 and January 1998. On the environmental profile and preferences of leading chemical exporters, see Suchanek 1997; Chemopetrol 1997; ACICR 1995.

15. ACICR 1998a.

16. ACICR 1997b.

17. Interview at ACICR, January 1998.

18. See environmental performance data in ACICR 1997a and ACICR 2002.

19. See environmental performance data in ACICR 1997a and ACICR 2002.

20. ACICR 1995; CEFIC 1996b; CEFIC 1997b; interview at ACICR, November 1997.

21. ACICR 1996, pp. 32, 33.

22. ACICR 2002; CEFIC 1998; European Dialogue 1998. Information on the ChemLeg and ChemFed initiatives is available at the web site of the European Commission, accessed in November 2002 via http://europe.eu.int. The web site of CEFIC provides a summary of the environmental cooperation and partnership activities of the association undertaken within the EU enlargement process, accessed in November 2002 via http://www.cefic.be.

23. ACICR 1998a.

24. ACICR 1998b, 1997b; interviews at ACICR, January 1998.

25. CEMC 1995, 1996, 1997, 1998a, 1999.

26. Interview with director of CBCSD, November 1997.

27. Interview at ACICR, November 1997.

28. ACICR 1997b, 1998a, 1998b; interview at ACICR regarding new Waste Management Act, November 1997.

29. Commission of the European Communities 2000.

30. Data for 2001 from PCCI 2002a. For similar data for 1996, see PCCI 1997a, p. i-3.

31. PCCI 1998, pp. 15, 16.

32. Data from 1995, reported in PCCI 1997a, p. vi-1.

33. Ibid.

34. PCCI 1997a.

35. Ibid.

36. Commission of the European Communities 2000.

37. PCCI 1997a, 1998; Commission of the European Communities 2000.

38. PCCI 1997a.

39. The PCCI was created in 1988 as a non-governmental organization representing the chemical industry. In 1992 it became a member of the Polish Chamber of Commerce and an associated member of CEFIC. See PCCI 1994.

40. PCCI 1994.

41. Lubiewa-Wielezynsky 1998.

42. Ibid.; PCCI 1997a.

43. PCCI 1994, 1997a, 1997b, 2002b.

44. PCCI 1994, p. 2.

45. CEFIC 1996a, 1997a; PCCI 1997a,b.

46. Interview at Industrial Chemistry Research Institute, March 1998. See also CEFIC 1997a.

47. CEFIC 1997a.

48. Responsible Care Program Poland, accessed via http://www.pipc.org.pl in November 2002.

49. Responsible Care Program Poland, accessed via http://www.pipc.org.pl in November 2002.

50. Barglik 1998; OECD 1999a,b; REC 1998a; interview with chairman of the Business Council for Sustainable Development (BCSD) of Poland, February 1998.

51. Barglik 1998; Klub Polskie Forum ISO 14,000 1996, 1997; World Environment Center 1999.

52. Barglik 1998; Klub Polskie Forum ISO 14,000 1997; interview with chairman of the BCSD of Poland, February 1998.

53. The membership of the Business Council for Sustainable Development of Poland is based on major multinational or export-oriented companies such as SGS, Cultur Holding, Universal, and Bertelsmann. Among the pioneers in adopting the ISO 14,000 standards in Poland are big multinational firms and exporters such as ABB ZAMECH, ABB Elta, ABB Elpar, AMICA Wronki, and Philips Lighting Poland. See Barglik 1998; Klub Polskie Forum ISO 14,000 1997; Lubiewa-Wielezynsky 1998; interview with chairman of BCSD of Poland, February 1998.

54. PCCI 1997a, p. ix-1.

55. PCCI web site, accessed in November 2002 via http://www.pipc.prg.pl.

56. Among the many international partners of the PCCI are Western chemical industry associations, CEFIC, UN, and EU agencies, and chemistry research institutes. See PCCI 1997a, p. ix-5.

57. Interviews at PCCI (February 1998) and at Industrial Chemistry Research Institute (March 1998).

58. PCCI 1998, p. 62. For a similar argument, see Lubiewa-Wieleczynsky 1998.

59. PCCI 1998, p. 62. Similar statements were made at an interview of a representative of the environmental department of the PCCI, February 1998.

60. PCCI 1998, pp. 62, 63.

61. Interview at department of environmental protection of PCCI, February 1998.

62. On the political and lobbying activities of the PCCI and its Advisory Group on Environment, see PCCI 1997a,b; interview at department for environmental protection of PCCI, February 1998.

63. PCCI 1998.

64. Commission of the European Communities 2000.

65. National Statistical Institute of the Republic of Bulgaria 1999.

66. BCCI 1998a.

67. Interviews with BCCI representatives, March 1998 and July 2000.

68. Danchev 1995.

69. Ibid.; Natov 1997.

70. Clean Industry Center 1998a; Bulgarian Industry Association 1996, 1997.

71. Clean Industry Center 1998a; interviews at Clean Industry Center, July 1998 and August 1998.

72. BCCI 1997; *European Dialogue* 1998.

73. BCCI 1997, 1998a; interview with BCCI representatives, March 1998.

74. The BCCI was created in 1994 as a part of the Bulgarian Industry Association to represent the political and economic interests of chemical enterprises. Its membership includes 65% of the chemical companies in Bulgaria and 85% of the people employed in the sector. See BCCI 1997.

75. Results of an enterprise-level survey as reported in BCCI 1998a.

76. BCCI 1997, 1998a, 1999; interviews with representatives of BCCI, March 1998 and February 1999.

77. BCCI 1998a, p. 152.

78. BCCI 1998a.

79. Interview with representatives of BCCI, March 1998 and February 1999. See also BCCI 1998a,b, 1999.

80. Interview with representatives of BCCI, March 1998; interview at department for Industrial Pollution and Waste Management of Ministry of the Environment and Waters of the Republic of Bulgaria, March 1998.

81. Interview with representatives of BCCI, February 1999.

82. BCCI 1997, 1998c, 1999; interview with representative of BCCI, February 1999.

83. Brickman, Jasanoff, and Illegen 1985, pp. 225, 226.

84. Garcia-Johnson 2000.

Chapter 2

1. REC 1996a, p. 11.

2. BCCI 1998a; REC 1996a.

3. ACICR 1998a; REC 1996a.

4. PCCI 1998; REC 1996a.

5. Ministry of Environment of Czech Republic 1991.

6. Ministry of Environment of Czech Republic 1995, 1996a.

7. Hereafter, in sections describing developments in the Czech Republic, I will use Ministry of Environment instead of Ministry of the Environment of the Czech Republic.

8. Cizkova and Orlikova 1997; Skalicky 1997.

9. Interviews at ACICR (January 1998), Department of Ecological Risks and Monitoring of the Ministry of the Environment of Czech Republic (November 1997), Ministry of Industry and Trade (November 1997), and with the Deputy Minister of Environment responsible for EU law approximation (November 1997). See also ACICR 1998a.

10. Ministry of Environment of Czech Republic 1991; REC 1995.

11. Statements by Vaclav Klaus in Green Circle 1996. See also Fagin and Jehlicka 1998.

12. REC 1995. Interviews with Czech environmental organizations: Green Circle, Greenpeace Czech Republic, Rainbow Movement Praha, and the REC-Czech Republic, November 1997.

13. Interviews with the Green Circle, the Czech Environmental Management Center, the Czech Business Council for Sustainable Development, and the Deputy Minister of Environment responsible for EU integration, November 1997.

14. Interviews with representatives of REC-Czech Republic, Green Circle, Greenpeace Praha, and Department of Ecological Risks and Monitoring of Ministry of Environment of Czech Republic, November 1997. See also Dujvelaar 1996; REC 1996b.

15. Interviews and e-mail communication with Department of Ecological Risks and Monitoring of Ministry of Environment of Czech Republic, November 1997 and March 1999.

16. ACICR 1998a; Ministry of Environment of Czech Republic 1997a; interview at Department of Ecological Risks and Monitoring of Ministry of Environment of Czech Republic, November 1997.

17. See ACICR 1997b, 1998b.

18. ACICR 1997b, 1998a.

19. Zakon 157 1998.

20. Zakon 352 1999.

21. See Zakon 157 1998; Zakon 352 1999; Blaha 2001. The complete list of governmental ordinances pertinent to the Act on Chemical Substances and Preparations is published at the web site of the Ministry of the Environment of the Czech Republic (http://www.env.cz).

22. Sir William Helcrow and Partners 1997; interviews at Department of International Relations of Ministry of Environment of Czech Republic, January 1998; interview with Deputy Minister of Environment responsible for EU integration, November 1997.

23. E-mail communication with the Department of Ecological Risks and Monitoring of Ministry of Environment of Czech Republic, March 1999. See also Blaha 2001.

24. Ministry of Environmental Protection, Natural Resources and Forestry of Republic of Poland 1991. See also Nowicki 1997; Sobolewski and Taylor 1996.

25. Ministry of Environment, Natural Resources and Forestry of Republic of Poland 1991.

26. Ministry of Health and Social Welfare of Republic of Poland 1996; Ministry of Environment, Natural Resources and Forestry of Republic of Poland 1997a.

27. See Ministry of Environment, Natural Resources and Forestry of Republic of Poland 1997a.

28. Hereafter the Ministry of Environment, Natural Resources and Forestry of the Republic of Poland will be referred to as the Ministry of Environment.

29. Interview with a representative of the Nofer Institute for Occupational Medicine responsible for drafting the law, August 1999, and at the Departments of International Relations and EU Integration of Ministry of Environment, Natural Resources and Forestry of Republic of Poland, February 1998.

30. Ministry of Environment, Natural Resources and Forestry of Republic of Poland 1997a; interview at Chancellery of Sejm, February 1998 and July 1999.

31. Interview at PCCI's department for environmental protection, February 1998.

32. PCCI 1998. Interviews at PCCI's department of environmental protection (February 1998) and with a representative of the Nofer Institution for Occupational Medicine (August 1999).

33. Andrijewski and Lewandowska 2002.

34. Communications with Chancellery of Sejm, July 1999 and June 2000.

35. Ministry of Environment, Natural Resources and Forestry of Republic of Poland 1998a; interview with representative of Nofer Institution for Occupational Medicine, August 1999.

36. Andrijewski and Lewandowska 2002; Statute on Chemical Substances and Preparations 2001.

37. Andrijewski and Lewandowska 2002.

38. Ibid.

39. Council of Ministers of Republic of Poland 2000.

40. The Ministry of Environment and Waters of the Republic of Bulgaria will hereafter be referred to as the Ministry of Environment.

41. World Bank 1994.

42. Interviews at Department of Industrial Pollution and Waste Management of Ministry of Environment and Waters of Republic of Bulgaria, March 1998.

43. BCCI 1998a.

44. BCCI 1997, 1998a, 1999; interview at Department of Industrial Pollution and Waste Management of Ministry of Environment and Waters of Republic of Bulgaria, March 1998.

45. Ministry of Environment and Waters of Republic of Bulgaria 1997a, Ministry of Health of Republic of Bulgaria 1997, and Ministry of Industry of Republic of Bulgaria 1997.

46. Ministry of Industry of Republic of Bulgaria 1997.

47. BCCI 1998c.

48. BCCI 1998a; interviews with BCCI representatives, March 1998 and February 1999.

49. Interviews with representatives of BCCI, February 1999 and July 2000.

50. National Assembly of Republic of Bulgaria 2000. See also Committee of Environment of National Assembly of Republic of Bulgaria 2000; interview with chairman of Committee on Environment of Bulgarian National Assembly, August 2000.

51. Zakon za Zashtita ot Vrednoto Vazdeistvie na Himicheskite Veshtestva, Preparati i Producti 2000.

52. Communication with Department for Industrial Pollution and Waste Management of Ministry of Environment, November 2002.

53. Alt and Gilligan 1994; Bates 1997; Berger 1996; Hall 1986; Garrett 1998; Garrett and Lange 1996; Gourevitch 1986; Katzenstein 1985; Weir and Skocpol 1985.

54. On the history of the Polish environmental movement, see Glinski 1996; Glinski 1998; Hicks 1996. On the development of Bulgarian environmental groups, see Baumgartl 1997; Vari et al. 1993. On the development of Czech environmental organizations, see Markova 1996; Dujvelaar 1996; Vari et al. 1993. See also REC 1995.

55. Interviews with representatives of environmental non-governmental organizations in Bulgaria, the Czech Republic, and Poland, including REC-Poland, the Institute for Sustainable Development (Poland), Social Ecology Institute (Poland), Deti Zeme Praha (The Czech Republic), Greenpeace Praha (The Czech Republic), REC-Czech Republic, Green Circle (The Czech Republic), REC-Bulgaria, Green Balkans (Bulgaria), Borrowed Nature (Bulgaria), and Za Zemjata (Bulgaria). See also REC 1996b.

56. Haas, Keohane, and Levy 1993; Keohane and Levy 1996.

57. *European Dialogue* 1998; *Enlarging the Environment* 1998.

58. *European Dialogue* 1998, p. 3.

Chapter 3

1. The Czech Republic was part of former Czechoslovakia (The Czech and Slovak Federal Republic) until January 1, 1993, when the two republics separated to become independent countries. This chapter focuses on the Czech Republic, but it refers to the former Czechoslovakia when analyzing political events that took place before 1993.

2. Martin, p. 1990; Pehe 1990; Vachudova 2001.

3. See Vachudova 2001 for a description of the broader influence of international commitments on the policies of the Civic Forum government.

4. World Bank 1992c.

5. Martin, p. 1990.

6. Data presented by Hertzman 1993, p. 25.

7. Federal Committee of the Environment of the Czech and Slovak Federal Republic 1992 and Hertzman 1993.

8. Ministry of the Environment of the Czech Republic 1991, p. 88.

9. Gallup Poll London 1990; Moldan 1997.

10. Poor electoral results were attributed to the lack of experience of the party, its inability to cooperate with the radical environmental movement, and the strong environmental platforms of the Civic Forum. See Jehlicka and Kostelecky 1992; Pehe 1990.

11. Fagin 1994; Fagin and Jehlicka 1998; Green Circle 1997; REC 1995; interviews with representatives of Green Circle, November 1997.

12. Federal Committee for the Environment 1992; Ministry of the Environment of the Czech Republic, 1991, 1995.

13. Ministry of the Environment of the Czech Republic 1991, p. 38.

14. Ibid., p. 36.

15. Interview with Bedrich Moldan, former Minister of the Environment of the Czech Republic, November 1997. See also Ministry of the Environment of the Czech Republic 1991.

16. Act on Clean Air of the Czech and Slovak Federal Republic (No. 309/1991).

17. Act on the Environment of the Czech and Slovak Federal Republic (No. 17/1992).

18. Federal Assembly of the Czech and Slovak Federal Republic 1991. See also Cole 1995; Levy 1993a.

19. Interview with Bedrich Moldan, former Minister of the Environment of the Czech Republic, with representatives of Green Circle, and with a close associate of the late Federal Minister Vavrousek, November 1997. See also Levy 1993a.

20. Federal Assembly of the Czech and Slovak Federal Republic 1991.

21. Federal Committee of the Environment of the Czech and Slovak Federal Republic 1992. Interviews with Bedrich Moldan (former Minister of the Environment of the Czech Republic) and with the former head of the environment department of the Czech Electricity Company (CEZ), November 1997.

22. Interviews with the Green Circle, the Rainbow Movement, Ministry of the Environment of the Czech Republic, former Minister of the Environment, and former head of the environment department of CEZ, November 1997. See also Jehlicka and Kostelecky 1992 for data on the support for environmental reforms across administrative regions of the Czech Republic as indicated by electoral polls and the relative performance of the Green Party in the local elections 1990.

23. Act on Clean Air of the Czech and Slovak Federal Republic (No. 309/1991). See also Cole 1995; Murley 1995.

24. Act of the Czech National Council on the State Administration of Air Protection and Charges for the Pollution of Air (No. 389/91).

25. Maillet 1994.

26. Ceska Inspekce Zivotniho Prostredi 1997; Ministry of the Environment of the Czech Republic 1996b.

27. Boehmer-Christiansen and Skea 1991; Liberatore 1992.

28. Duina 1999.

29. CEZ 1999, 2002; International Energy Agency 1992, 1994a.

30. CEZ 1999, p. 42.

31. CEZ 1997, 1998, 1999.

32. CEZ 1997, 1998, 1999.

33. Vlacek 1992.

34. Andrews 1993; Cole 1995; Fagin 1994; Maillet 1994.

35. Vlacek 1992; Federal Ministry of the Economy of the Czech and Slovak Federal Republic 1991.

36. Levy 1993a; Pehe 1990.

37. Interviews with a former head of the environmental department of CEZ, November 1997.

38. International Energy Agency 1992; Pehe 1990.

39. Vlacek 1992. Interview with a representative of SEVEN, November 1997.

40. International Energy Agency 1992; World Bank 1992a.

41. International Energy Agency 1992. Interview with a representative of SEVEN, November 1997.

42. Axelrod 1999; Connolly and List 1996; Pehe 1990.

43. Ministry of the Environment of the Czech Republic 1991, p. 39.

44. Interviews with representatives of the Program for Energy Efficiency, Rainbow Movement Praha, Greenpeace Praha, and SEVEN, November 1997.

45. Interview with a representative of the Program for Energy Efficiency, November 1997.

46. Vlacek 1992; World Bank 1992b.

47. Vlacek 1992.

48. World Bank 1992b.

49. CEZ 1999, p. 28.

50. Fagin and Jehlicka 1998. On the structure of Czech external borrowing, see CEZ 1998, 1999.

51. CEZ 1997, p. 61.

52. Data from 1998, provided by CEZ 1999.

53. CEZ 1999.

54. Ibid., p. 28.

55. International Energy Agency 2001.

56. CEZ 1999, p. 28.

57. Ibid., p. 10.

58. Sir William Halcrow and Partners, Ltd. 1997. See also Bizek 2001.

59. Interviews at Department for Air Protection of the Czech Ministry of the Environment. See also Bizek 2001.

60. Axelrod 1999; Fagin and Jehlicka 1998.

Chapter 4

1. Nowicki 1997; State Inspectorate of Environmental Protection of the Republic of Poland 1993.

2. Millard 1999; Sachs 1994.

3. Hicks 1996; Jendroska 1998.

4. For detailed histories of the development of the independent ecological movement in during the communist regime, see Glinski 1996, 1998; Hicks 1996. See also Kabala 1993; Cole 1998; Andersson 1999.

5. Andersson 1999; Karaczun 1993a.

6. Karaczun 1993a, p. 6.

7. Ministry of Environment, Natural Resources and Forestry of the Republic of Poland 1991.

8. Anderson 1999; Cole 1998; Ministry of Environment, Natural Resources and Forestry of the Republic of Poland 1991; Nowicki 1997; Sobolewski et al. 1996; World Bank 1991, 1992e.

9. In 1990, there were 49 administrative regions (wojewodztwa) in Poland. After the administrative reforms of 1999, the number of administrative regions was reduced to 16.

10. Brown et al. 1998; Ministry of Environment, Natural Resources and Forestry of the Republic of Poland 1991.

11. Act on the State Inspectorate of Environmental Protection of the Republic of Poland 1991.

12. Communication with the PPGC, November 2002.

13. McNicholas and Speck 2000, table 4. Similar data is quoted by Anderson and Fiedor 1997, p. 206.

14. Anderson and Fiedor 1997; Ministry of Environment, Natural Resources and Forestry of the Republic of Poland 1995; McNicolas and Speck 2000; Nowicki 1997; Zylicz 1993.

15. Ministry of Environment, Natural Resources and Forestry of the Republic of Poland 1997a.

16. OECD 1998.

17. Bochniarz and Bolan 1998; Council of Ministers of the Republic of Poland 2000; GUS 1999; National Fund for Environmental Protection and Water Management of the Republic of Poland 1999; Nowicki 1997; OECD 1998.

18. REC 1993.

19. Interview with the Director of the Environmental Policy Department of the Ministry of Environment, Natural Resources and Forestry of the Republic of Poland, February 1998.

20. Millard 1999; Taras 1998; World Economy Research Institute 2000.

21. Andersson 1999; Karaczun 1993a, b; Karaczun 1996.

22. Andersson 1999; Glinski 1998; Karaczun 1993b; Karaczun 1996; Millard 1999; interviews with representatives of environmental groups, February 1998.

23. Act on Environmental Protection and Management of the Republic of Poland.

24. Karaczun 1993b; Millard 1999.

25. EkoFinanse 1997.

26. Council of Ministers of the Republic of Poland 2000.

27. World Bank 1997; World Economy Research Institute 2000.

28. Polish Environmental Law Association 1997. Interviews with representatives of the Institute for Sustainable Development, Polish Ecological Club, and Social Ecological Institute, February 1998.

29. Balcerowicz 1997; Unia Wolnosci 1997a,b; interviews with members of the Forum of Ecological Leaders of Unia Wolnosci (Freedom Union), and interviews with representatives of the Institute for Sustainable Development and Social Ecological Institute, February 1998.

30. Ministry of Environment, Natural Resources and Forestry of the Republic of Poland 1990.

31. Karaczun 1993a; Nowicki 1997; State Inspectorate of Environmental Protection of the Republic of Poland 1993.

32. OECD 1995, p. 33. See also State Inspectorate of Environmental Protection of the Republic of Poland 1993.

33. Hertzman 1993; Karaczun 1993a.

34. Karaczun 1993a.

35. Levy 1993, p. 92.

36. New sources were defined as installations, the construction of which began after 1991, when the ordinance came into force, or start operation in 1995. See Ministry of Environment, Natural Resources and Forestry of the Republic of Poland 1990.

37. Ministry of Environment, Natural Resources and Forestry of the Republic of Poland 1990.

38. European Council 1988; Ministry of Environment, Natural Resources and Forestry of the Republic of Poland 1990; OECD 1995. For a graphic comparison between Polish standards and EU and LRTAP requirements, see Jankowski et al. 1998, annex I, p. 11.

39. Karaczun 1996; World Bank 1997, 1998.

40. *Financial Times* 1998; Jankowski et all. 1998; Krakow Academy of Economics 1996; World Bank 1997, 1998.

41. Data for 1996, provided by State Inspectorate of Environmental Protection of the Republic of Poland 1998, p. 28.

42. International Energy Agency 1990.

43. International Energy Agency 1990, 1994b; PPGC 1999a.

44. International Energy Agency 1994b; interviews at PPGC, February 1998 and June 2000.

45. Balcerowicz 1999; Buczkowski 2000; Biuletyn Energetyczny 1999; Duda 1999.

46. International Energy Agency 1990, p. 77.

47. PPGC 1997, 1999b.

48. Hicks 1996.

49. The cost estimates for the country as a whole are provided by Council of Ministers of the Republic of Poland 2000. Industry estimates are provided in World Bank 1998. For other assessments of the cost of compliance with air emis-

sion standards according to different compliance scenarios, see Krakow Academy of Economics 1996; Jankowski et al. 1998; EkoFinanse 1997. For a discussion of the cost implications of emission reduction requirement for individual power plants, see Biuletyn Energeticzny 1999.

50. Ministry of the Economy of the Republic of Poland 1997, 2002.

51. Interviews with representatives of the PPGC and the Energy Department of the Ministry of the Economy, February 1998.

52. Interview with a representative of the Energy Department of the Ministry of the Economy, February 1998.

53. Data for 1997 from GUS 1999.

54. International Energy Agency 1994b; Institute for Sustainable Development 1999; Karaczun 1993b; interviews with representatives of PPGC and Institute for Sustainable Development, June 2000.

55. Biuletyn Energetyczny 2000, p. 3. See also Biuletyn Energetyczny 1999, 2000; Ministry of Environment, Natural Resources and Forestry of the Republic of Poland 2000; Tokarczuk 2000.

56. Ministry of Environment, Natural Resources and Forestry of the Republic of Poland 1998d; Tokarczuk 2000; Zylicz 1998, interviews at Department for International Relations of the Ministry of Environment, Natural Resources and Forestry of the Republic of Poland and the Council of Ministers, June 2000.

57. Tokarczuk 2000, p. 9.

58. Karaczun 1996, p. 53.

59. Interview with representative of PPGC, February 1998.

60. Ibid.

61. Andersson 1999; *Biuletyn Energetyczny* 2000; Krakow Academy of Economics 1996; Zylicz 1998; interviews with economic advisers to the Ministry of Environment, Natural Resources and Forestry of the Republic of Poland, February 1998; interview with a representative of the PPGC, February 1998.

62. For a detailed discussion of the role of economic assessments in facilitating a political agreement on the adoption of international air pollution standards, see Botcheva 2001.

63. Biuletyn Energeticzny 2000; Nowicki 1997; OECD 1995; PPGC 2000.

64. See also GUS 1997, 1999.

65. Council of Minister of the Republic of Poland 2000.

66. National Fund for Environmental Protection and Water Management 1999 and Biuletyn Energetyczny 1999. Interviews with representatives of the PPGC, the Air Protection Department of the Ministry of Environment, Natural Resources and Forestry of the Republic of Poland, economic advisers of the Ministry of the Environment, and with the Director of the Air Protection Unit of the National Fund for Environmental Protection and Water Management, February 1998 and June 2000.

67. Ministry of Environment, Natural Resources and Forestry of the Republic of Poland, Ministry of Industry and Trade, and Ministry of Regional Planning and Construction of the Republic of Poland 1996.

68. Ministry of Environment, Natural Resources and Forestry of the Republic of Poland and Ministry of Industry and Trade of the Republic of Poland 1996.

69. Interviews with representative of the PPGC and the Energy Department of the Ministry of the Economy, February 1998. See also Duda 1999; Steinoff 1999.

70. Ministry of Environment, Natural Resources and Forestry of the Republic of Poland 1997b, p. 42.

71. Ministry of Environment, Natural Resources and Forestry of the Republic of Poland 1998b, c.

72. Interview with Tomas Zylicz, June 2000.

73. European Council 1988; Ministry of Environment, Natural Resources and Forestry of the Republic of Poland 1998c.

74. European Council 1988; Ministry of Environment, Natural Resources and Forestry of the Republic of Poland 1998c.

75. Interview a representative of the PPGC, June 2000.

76. Commission of the European Communities 1997b; Council of Minister of the Republic of Poland 2000.

Chapter 5

1. Baumgartl 1997, p. 62. On Ekoglasnost and its role in the democratic changes in 1989, see Baumgartl 1993; Botcheva 1996; Fisher 1993.

2. Kalinova and Baeva 2000.

3. Baumgratl 1997; Fisher 1993; interview with former member of Ekoglanost, August 2000.

4. Fisher 1993; Karasimeonov 1997.

5. Baumgartl 1997; Penchovska 1997; REC 1995; interviews with Za Zemjata, REC-Bulgaria, Green Balkans, Infomacionen Centar za Ekologia, and former members of Ecoglasnost, March 1998 and August 2000.

6. Baumgartl 1997; Bochniarz and Georgieva 1992; Koulov 1998.

7. Dimova 1995, p. 3.

8. Bochniarz and Georgieva 1992; Georgieva 1993; Koulov 1998; Yarnal 1995.

9. Daskalov 1998; Kalinova and Baeva 2000.

10. Georgieva and Moore 1997; Yarnal 1995.

11. World Bank 1994.

12. Ministry of the Environment and Waters of the Republic of Bulgaria 1997b; REC 1995, p. 41.

13. REC 1995; interview with Green Balkans, Za Zemiata, and Informacionen Centar po Ekologia, July 1997 and July 2000.

14. Interview with Green Balkans, July 1997.

15. Interviews with Za Zemjata, Ekoglasnost, Green Balkans, REC-Bulgaria, and the chairman of the Parliamentary Committee of the Environment, March 1998 and July–August 2000.

16. Ministry of the Environment and Waters of the Republic of Bulgaria 1997b and interviews with Green Balkans, Za Zemiata, and Informacionen Centar po Ekologia, July 2000.

17. National Center for Environment and Sustainable Development 1990–2001.

18. Black Sea Energy Research Center 1996; NEK 1996, 1998, 1999.

19. World Bank 1992d.

20. NEK 2000.

21. Committee of Energy of the Republic of Bulgaria 1998; Zakon za Energetikata i Energiinata Effectivnost 1999.

22. Ministry of Energy and Energy Resources of the Republic of Bulgaria 2002. On the course and environmental implications of energy reforms in Bulgaria, see Andonova 2002.

23. Energoproekt 1996, table 1.4-1; Kikuchi 1997; National Center for Environment and Sustainable Development 1990–1999; OECD 1996.

24. Energoproekt 1996, pp. 83–86.

25. Ministry of the Environment and Waters of the Republic of Bulgaria 1998c, table D.

26. Komitet po Associirane Bulgaria-EC 2000; Ministry of the Environment and Waters of the Republic of Bulgaria 1999.

27. Interviews with representatives of Energoproekt, the Committee of Energy, NEK, March 1998 and July 2000.

28. Interview with a representative of NEK, July 2000.

29. Matev and Nivov 1997; OECD 1996.

30. Committee of Energy of the Republic of Bulgaria 1999.

31. National Report of the Republic of Bulgaria the Reduction of SO_2 and Dust Emissions 1998.

32. Koulov 1998.

33. National Center for Environment and Sustainable Development 1995; OECD 1996; Yarnal 1995.

34. Baumgartl 1993; Energoproekt 1996; Hertzman 1993; National Center for Environment and Sustainable Development 1990–1999; National Report of the Republic of Bulgaria on the Reduction of SO_2 and Dust Emissions 1998; UNECE 1993.

35. Kikuchi 1997.

36. Hertzman 1993.

37. UNECE 1993.

38. EBRD 1992; OECD 1996; UNECE 1993; World Bank 1994.

39. EBRD 1992.

40. Ministry of the Environment and Waters of the Republic of Bulgaria 1991.

41. OECD 1996, pp. 41, 48.

42. EBRD 1992; interview with director of legal department of Ministry of the Environment and Waters of the Republic of Bulgaria, June 1997.

43. Interviews with members of Ekoglasnost, Za Zemjata, and with Infomacionen Centar za Ekologia, March 1998 and July 2000.

44. Interviews with former officials from the Ministry of the Environment and Waters of the Republic of Bulgaria, March 1998.

45. Council of Ministers of the Republic of Bulgaria 1995.

46. National Assembly of the Republic of Bulgaria 1996a.

47. Ibid., article 10.

48. National Assembly of the Republic of Bulgaria 1996a, p. 14.

49. Ibid., pp. 14–15.

50. National Assembly of the Republic of Bulgaria 1996a,b.

51. National Assembly of the Republic of Bulgaria 1996b.

52. Zakon za Chistotata na Atmosfernia Vazduh 1996.

53. World Bank 1996.

54. Interviews with former officials from the Ministry of the Environment and with representatives of the NEK, March 1998.

55. Commission of the European Communities 1997a, pp. 83, 84.

56. Ministry of the Environment and Waters of the Republic of Bulgaria 2001.

57. Interviews at Air Protection Department of Ministry of Environment and Waters of the Republic of Bulgaria, March 1998 and July 2000. See also Ministry of the Environment and Waters of the Republic of Bulgaria 1998a and Ministry of the Environment and Waters of the Republic of Bulgaria 1998b. A list of the international projects of Ministry of the Environment and Waters of the Republic of Bulgaria was made available at http://www.moew.government.bg.

58. Ministry of the Environment and Waters of the Republic of Bulgaria 1998b, 1998c, 1999a.

59. Ministry of the Environment and Waters of the Republic of Bulgaria 2000.

60. Ministry of the Environment and Waters of the Republic of Bulgaria, Ministry of Health, Ministry of Industry, and Ministry of Regional Development 1998, 1999.

61. Ministry of the Environment and Waters of the Republic of Bulgaria 1998d.

62. Committee of Energy of the Republic of Bulgaria 1998, 1999.

63. Ministry of the Environment and Waters of the Republic of Bulgaria 2000. On the high level of formal harmonization of the EU air pollution Acquis in Bulgaria, see Ministry of the Environment and Waters of the Republic of Bulgaria 2001; UNECE 2000.

Conclusion

1. Rogowski 1989; Frieden 1991; Milner 1988; Alt 1987; Bates 1997; Garrett 1998; Goldstein 1986; Gourevitch 1986; Hall 1986; Katzenstein 1978; Rodrik 1999; Rogowski 1987; Berger 1996; Keohane and Milner 1996.

2. Heritier et al. 2001; Cowles; Caporaso and Risse 2001; Borzel and Risse 2000; Borzel 2000; Duina 1999; Haverland 2000; Haas 1998; Haigh 1990; Glachant 2001; Knill and Lehmkuhl 1999; Grabbe 2001.

3. Important recent contributions of two-level analyses in environmental politics include the following: Darst 2001; DeSombre 2000; Garcia-Johnson 2000; Keck and Sikkink 1998; O'Neill 2000; Steinberg 2002; Young 2002.

4. Prominent examples: Dawisha and Parrot 1997; Ekiert 1996; Fish 1998; Haggard and Kaufman 1995; Hellman 1998; Kitschelt et al. 1999b; Klarer and Moldan 1997; Przeworski 1991; Shugart and Carey 1992; Stark and Bruszt 1997; Vari and Tamas 1994.

5. A new generation of research exemplifies the effort to understand the significance of external factors on policy choices of transition states. See e.g. Stone 2002; Kelley 2001; Sissenich 2002.

6. Garcia-Jonhson 2000.

7. Vogel 1995, 1997.

8. Ruggie 2002.

9. For a detailed review of contributions to this perspective, see Martin and Simmons 1998.

10. Botcheva and Martin 2001; Evans, Jacobson, and Putnam 1993; Duffield 1991; Gurowitz 1999; Haas 1998; Haggard and Simmons 1987; Kelley 2001; Martin and Sikkink 1993; Moravcsik 1995; Simmons 1998.

11. For a review, see Ruggie 1998a,b.

12. See Checkel 2001; Klotz 1995; Katzenstein 1996; Haas 1989; Finnemore 1996; Keck and Sikkink 1998; Risse, Ropp, and Sikkink 1999.

13. Choucri 1993; Haas, Keohane, and Levy 1993; Victor et al. 1998; Brown, Weiss, and Jacobson 1998; Mitchell 1994; Young 1989b, 1994, 1999; Young and Osherenko 1993; Litfin 1998; Lipschutz 1996; Wapner 1995; Keck and Sikkink 1998.

14. Haas 1989; Risse-Kappen 1995; Checkel 1997; Eising 2002; Schimmelfenning 2001; Young 2002.

15. Haas 1989; Checkel 1997; Risse-Kappen 1995.

16. Haas 1968; Keohane and Nye 1972.

17. Tallberg 2002; Burley and Mattli 1993; Grant, Matthews, and Newell 2000.

18. See Ruggie 2002 on a similar note.

19. The literature on globalization and governance is vast. For recent comprehensive contributions that also review that literature, see Held 1999; Giddens 2000; Nye and Donahue 2000; Mol 2001; Keohane and Nye 2000; Lofdahl, C. 2002; Stiglitz 2002.

20. See e.g. Daly 1995; Costanza 1995; Environmental Defense Fund 1999.

21. See e.g. Bhagwati 1993; Bhagwati and Hudec 1996; World Bank 1992.

22. Vogel 1995, 1997; Eliste and Fredriksson 1998; Bhagwati and Hudec 1996; Zarrilli et al. 1997; Garcia-Johnson 2000; Mol 2001.

23. This presumption is also squarely challenged by a recent study of biodiversity protection in Costa Rica and Bolivia (Steinberg 2002).

Bibliography

ACICR (Association of the Chemical Industry of the Czech Republic). 1994. The Association of the Chemical Industry of the Czech Republic (brochure).

ACICR. 1995. Responsible Care: Health, Safety, and Environment Protection.

ACICR. 1996. Annual Report 1995.

ACICR. 1997a. Annual Report 1996.

ACICR. 1997b. Standpoint of the Association of the Chemical Industry of the Czech Republic with Respect to the Draft Bill on Chemical Substance and Preparations.

ACICR. 1998a. Impact of the Commission's White Paper on the Chemical Industry in CEE Countries. Country Report: The Czech Republic. Brussels: CEFIC-PHARE.

ACICR. 1998b. Communication to Regional Environmental Center.

ACICR. 2002. Annual Report 2001.

Act of the Czech National Council on the State Administration of Air Protection and Charges for the Pollution of Air, No. 389/91, September 10, 1991. English translation.

Act on Clean Air of the Czech and Slovak Federal Republic, No. 309/1991. English translation.

Act on the Environment of the Czech and Slovak Federal Republic. No. 17/1992. English translation.

Act on Environmental Protection and Management of the Republic of Poland. 1980. English translation.

Act on the State Inspectorate of Environmental Protection of the Republic of Poland. 1991. English translation.

Alt, J. 1987. "Crude Politics: Oil and the Political Economy of Unemployment in Britain and Norway, 1970–85." *Journal of Political Science* 17: 149–199.

Alt, J., and Gilligan, M. 1994. "The Political Economy of Trading States: Factor Specificity, Collective Action Problems, and Domestic Political Institutions." *Journal of Political Philosophy* 2, no. 2: 165–192.

Anderson, G., and Fiedor, B. 1997. "Environmental Charges in Poland." In *Controlling Pollution in Transition Economies*, ed. R. Bluffstone and B. Larson. Elgar.

Andersson, M. 1999. *Change and Continuity in Poland's Environmental Policy*. Kluwer.

Andonova, L. 2002. "The Challenges and Opportunities for Reforming Bulgaria's Energy Sector." *Environment* 44, no. 10: 8–19.

Andrews, R. 1993. "Environmental Policy in the Czech and Slovak Republic." In *Environment and Democratic Transition*, ed. A. Vari et al. Kluwer.

Andrijewski, M., and Lewandowska, A. 2002. "Chemical Control Regulations In Poland." Presentation of Bureau for Chemical Substances and Preparations. Delivered at Conference on Chemical Safety, Portorose, Slovenia.

Axelrod, R. 1999. "Democracy and Nuclear Power in the Czech Republic." In *The Global Environment*, ed. R. Axelrod and N. Vig. CQ Press.

Baker, S., and Jehlicka, P., eds. 1998. Dilemmas of Transition: The Environment, Democracy and Economic Reform in East Central Europe. Special Issue of *Environmental Politics* 7:1.

Balcerowicz, L. 1997. "Unia Wolnosci—Dlaczego Ekolodzy?" *Za Zjelona Alternatywa*, no. 18.

Balcerowicz, L. 1999. "First—Privatization and Restructuring." *Biuletyn Energetyczny*, June: 7.

Barglik, J. 1998. "State of Environmental Management in Poland." Central and Eastern Europe Business and Industry NGOs' Contribution to Environment for Europe Conference, Aarhus, Denmark.

Bates, R. 1997. *Open Economy Politics: The Political Economy of the World Coffee Trade*. Princeton University Press.

Baumgartl, B. 1993. Environmental Protest as a Vehicle of Transition: The Case of Ekoglasnost in Bulgaria." In *Environment and Democratic Transition*, ed. A. Vari et al. Kluwer.

Baumgartl, B. 1997. *Transition and Sustainability: Actors and Interests in Eastern European Environmental Policies*. Kluwer Law International.

Baun, M. 2000. *A Wider Europe: The Process and Politics of European Union Enlargement*. Rowman & Littlefield.

BCCI (Bulgarian Chamber of the Chemical Industry). 1997. Otcheten Doklad na Upravitelnia Savet pred Tretoto Otchetno Obshto Sabranie. Report to Management Council of the BCCI.

BCCI. 1998a. Impact of the Commission's White Paper on the Chemical Industry in CEE Countries. Country Report: Bulgaria. Brussels: CEFIC-PHARE.

BCCI. 1998b. Osnovni Nasoki 1998–2000. Report to Management Council of the BCCI.

BCCI. 1998c. "Basic Principles and Contents of the Law on Chemical Substances." In BCCI, Impact of the Commission's White Paper on the Chemical Industry in CEE Countries. Country Report: Bulgaria. Brussels: CEFIC-PHARE.

BCCI. 1999. Doklad ot Predsedatelja na BCHP do Izpalnitelnite Direktori na Targovski Druzesva ot Himicheskata Promishlenost. Report of the Chairman of the BCCI to Managing Directors of Chemical Companies.

Becker, G. 1983. "A Theory of Competition among Pressure Groups for Political Influence." *Quarterly Journal of Economics* 98: 371–400.

Beckman, A. 2002. "Pushing on the Door: The Role of Central and East European NGOs in Enlarging the EU." In *EU Enlargement and Environmental Quality*, ed. S. Crisen and J. Carmin. Woodrow Wilson International Center.

Bell, R.G. 2002. "EU Membership: Boon or Bane for the Environmental Community in the Accession Countries?" In *EU Enlargement and Environmental Quality*, ed. S. Crisen and J. Carmin. Woodrow Wilson International Center.

Berger, S., and Dore, R., eds. 1996. *National Diversity and Global Capitalism*. Cornell University Press.

Bhagwati, J. 1993. "The Case for Free Trade." *Scientific American*, November: 42–49.

Bhagwati, J., and Hudec, R., eds. 1996. *Fair Trade and Harmonization*. MIT Press.

Biuletyn Energetyczny. 1999. "Strategy and Money." June: 33.

Biuletyn Energetyczny. 2000. "Modern Pro-Ecological Solution." June: 22–23.

Bizek, V. 2001. "Implementation and Enforcement of the Directives on Fuel Quality, Vehicle Emissions and Air Quality in the Czech Republic." Presentation at conference on Environmental Approximation in the Western Newly Independent States, March 22–23.

Black Sea Energy Research Center. 1996. Black Sea Energy Review: Bulgaria. Final Report.

Blaha, Karel. 2001. "Chemical Legislation in the Czech Republic: Two Years of Implementation." Presentation at workshop on Application of EU Chemical Legislation in the Candidate Countries: Progressing towards EU Enlargement, Istanbul.

Bochniarz, A., and Bolan, R. 1998. "Sustainable Institutional Design in Poland: Putting Environmental Protection on a Self-Financing Basis." In *Environmental Protection in Transition.*, ed. J. Clark and D. Cole. Ashgate.

Bochniarz, Z., and Georgieva, K. 1992. "Environmental Protection and Economic Development in Bulgaria: Legacy, Actions, and Prospects." In *Environment and Development in Bulgaria*. Papers from the International Workshop on Institutional Design for Environmental Protection in Bulgaria, Shtarkelovo Gnezdo.

Boehmer-Christiansen, S., and Skea, J. 1991. *Acid Politics: Environmental and Energy Policies in Britain and Germany*. Belhaven.

Bofinger, P. 1998. "The Political Economy of Eastern Enlargement." In *Forging an Integrated Europe*, ed. B. Eichengreen and J. Frieden. University of Michigan Press.

Borzel, T. 2000. "Why There Is No 'Southern Problem.' On Environmental Leaders and Laggards in the European Union." *Journal of European Public Policy* 7, no. 1: 141–162.

Borzel, T., and Risse, T. 2000. "When Europe Hits Home: Europeanization and Domestic Change." Paper presented at annual convention of American Political Science Association, Washington.

Botcheva, L. 1996. "Focus and Effectiveness of Environmental Activism in Eastern Europe." *Journal of Environment and Development* 5, no. 3: 292–308.

Botcheva, L. 2001. "Expertise and International Governance: Eastern Europe and the Adoption of European Union Environmental Legislation." *Global Governance* 7: 197–224.

Botcheva, L., and Martin, L. 2001. "Institutional Effects on State Behavior: Convergence and Divergence." *International Studies Quarterly* 45: 1–26.

Brickman, R., Jasanoff, S., and Ilgen, T. 1985. *Controlling Chemicals. The Politics of Regulation in Europe and the United States*. Cornell University Press.

Brown, H., et al. 1998. "Environmental Reforms in Poland: A Case for Cautious Optimism." *Environment* 40, no. 1: 10–13.

Brown Weiss, E., and Jacobson, K. 1998. *Engaging Countries. Strengthening Compliance with International Environmental Accords*. MIT Press.

Buczkowski, J. 2000. "Privatization of the Sector," *Biuletyn Energetyczny*, June: 15.

Bulgarian Industry Association. 1996. Centar za Chista Industria.

Bulgarian Industry Association. 1997. The Clean Industry Center at the Bulgarian Industrial Association.

Burley, A., and Mattli, W. 1993. "Europe before the Court: A Political Theory of Legal Integration." *International Organization* 47, no. 1: 41–76.

Caddy, J. 1997. "Hollow Harmonization? Closing the Implementation Gap in Central European Environmental Policy." *European Environment* 7: 73–39.

Calow, P. 1997. *Controlling Environmental Risks from Chemicals. Principles and Practice*. Wiley.

Cameron, D. 1978. "The Expansion of the Public Economy: A Comparative Analysis." *American Political Science Review* 72, no. 4: 1243–1261.

Cameron, D. 2002. "The Challenges of EU Accession." Paper presented at annual meeting of American Political Science Association, Boston.

Carius, A. 2002. "Challenges for Governance in a Pan-European Environment: Transborder Cooperation and Institutional Coordination." In *EU Enlargement and Environmental Quality*, ed. S. Crisen and J. Carmin. Woodrow Wilson International Center.

CBCSD (Czech Business Council for Sustainable Development). 1997. Czech Business Council for Sustainable Development. Prague.

CEFIC (European Chemical Industry Council). 1992. Central and Eastern Europe. The Challenges, Opportunities, and Risks for the European Chemical Industry. Industry Memo.

CEFIC. 1996a. *CEFIC Responsible Care 1996.*

CEFIC. 1996b. CEFIC Position on the Trade and Economic Aspects of EU Enlargement: Opening of the EU to Central and East European Countries. Position paper.

CEFIC. 1997a. Responsible Care. Annual Report 1997.

CEFIC. 1997b. Annual Report, 1996.

CEFIC. 1998. Impact of the Commission's White Paper on the Chemical Industry in CEE Countries.

CEMC (Czech Environmental Management Center). 1995. Annual Report 1994.

CEMC. 1996. "The Case of the Czech Republic: Towards the Dialogue on Sustainability via Establishment of Environmentally Oriented Business Structure." Paper presented at International Conference on The Integration of Environmental Policies and Strategies for Economic Reform in the NIS and Central Europe, Finland.

CEMC. 1997. Priprave Ceskych Podniku na Harmonizaci s Legislativou Evropske Unie. Project document.

CEMC. 1998a. Activity a Zamery Svazy Prumyslu a Dopravy CR a Ceske Podnikatelske Rady pro Udrzitelny Rozvoj Zamerene na Pripojeni CR k EU. Policy paper.

CEMC. 1998b. Central and Eastern Europe Business and Industry NGOs: Contribution to Environment for Europe Conference, Aarhus, Denmark.

CEMC. 1999. CENTICA. Koordinacni a Informacni Stredisko pro Pripravu Prumyslu na Pripojeni k EU v Oblasti Environmentalni Legislativy a Politiky.

Ceska Inspekce Zivotniho Prostredi (Environmental Inspectorare of Czech Republic). 1997. Balance Emisi Znecistujcich Latek v Roce 1996.

CEZ (Czech Electricity Company). 1997. Annual Report 1996. Prague: Kuklik.

CEZ. 1998. Annual Report 1997. Prague: Kuklik.

CEZ. 1999. Annual Report 1998. Prague: Business Information Group.

CEZ. 2002. CEZ and the Environment.

Checkel, J. 1997. "International Norms and Domestic Politics. Bridging the Rationalist-Constructivist Divide." *European Journal of International Relations* 3, no. 4: 473–496.

Checkel, J. 2001. "Why Comply? Social Learning and European Identity Change." *International Organization* 55, no. 3: 553–588.

Chemopetrol. 1997. *Environmental Protection Report.* Litvinov.

Choucri, N. 1993. *Global Accords. Environmental Challenges and International Responses*. MIT Press.

Churchill, R., et al. 1995. "The 1994 UN ECE Sulphur Protocol." *Journal of Environmental Law* 7, no. 2: 169–199.

Cizkova, H., and Orlikova, A. 1997. "Environmental Approximation Experience from the Czech Republic." Presentation at Green Globe Europe Conference, The Hague.

Clapp, J. 1998. "The Privatization of Global Environmental Governance: ISO 1400 and the Developing World." *Global Governance* 4: 295–316.

Clean Industry Center. 1998a. "Report on the State of Environmental Management in Bulgaria." Central and Eastern Europe Business and Industry NGOs' Contribution to Environment for Europe Conference, Aarhus, Denmark.

Clean Industry Center. 1998b. Illustrative Report to the Survey Examining the Determinants of Firm Level Environmental Performance in Bulgarian Industry.

Cole, D. 1998. *Instituting Environmental Protection. From Red to Green in Poland*. St. Martin's Press.

Cole, J. 1995. "European Union Air Protection Legislation: Is the Czech Republic Ready?" *European Environment* 5: 145–149.

Commission of the European Communities. 1996a. *European Community Environmental Legislation*, volume 2: *Air*. Luxembourg: Office for Official Publications of European Communities.

Commission of the European Communities. 1996b. *European Community Environmental Legislation. Volume. 3. Chemicals, Industrial Risks and Biotechnology*.

Commission of the European Communities. 1997a. *Agenda 21. Commission Opinion on Bulgaria's Application for Membership to the European Union*.

Commission of the European Communities. 1997b. *Agenda 21. Commission Opinion on Poland's Application for Membership of the European Union*.

Commission of the European Communities. 1997c. Communication to Council and the European Parliament on a Community Strategy to Combat Acidification. Document 97/105.

Commission of the European Communities. 1997d. *Agenda 21. Strengthening the Union and Preparing for Enlargement*.

Commission of the European Communities. 1999a. Bulgaria. Results of the Bilateral Screening, 5–9 July. Chapter 22: Environment. Draft report.

Commission of the European Communities. 1999b. Regular Report from the Commission on Progress towards Accession. Final Report, October 13.

Commission of the European Communities. 1999d. Guide to the Electricity Directive.

Commission of the European Communities. 2000. Competitiveness of the Chemical Industry Sector in the CEE Candidate Countries. Report of the Enterprise Directorate General.

Commission of the European Communities. 2002. Towards the Enlarged Union.

Committee of Energy of the Republic of Bulgaria. 1998. Nacionalna Strategia za Razvitie na Energetikata i Energiinata Effektivnost do 2010 Godina. Policy document.

Committee of Energy of the Republic of Bulgaria. 1999. Action Plan for Meeting the Commitments of Republic of Bulgaria to International Environmental Agreements on the Basis of the National Energy Strategy up to 2010.

Committee of the Environment of the National Assembly of the Republic of Bulgaria. 2000. Tablica na Saotvetsvieto: Zakon za Himicheskite Veshtestva i Preparati.

Connolly, B., and List, M. 1996. "Nuclear Safety in Eastern Europe and the Former Soviet Union." In *Institutions for Environmental Aid*, ed. R. Keohane and M. Levy. MIT Press.

Costanza, R., et al. 1995. "A New Paradigm for World Welfare." *Environment* 37, no. 5: 17–44.

Council of Ministers of the Republic of Bulgaria. 1995. Reshenie 434, October 23, 1995. Official copy provided by library of National Assembly of Bulgaria.

Council of Ministers of the Republic of Bulgaria. 1999. *Doklad za Sastoianieto na Okolnata Sreda prez 1998. Zelena Kniga.*

Council of Ministers of the Republic of Bulgaria. 2000. *Doklad za Sastoianieto na Okolnata Sreda prez 1999. Zelena Kniga.*

Council of Ministers of the Republic of Poland. 2000. Poland's Position Paper in the Area of Environment for the Accession Negotiation with the European Union.

Cowles, M.G. 2001. "The Transatlantic Business Dialogue and Domestic Business Government Relations." In *Transforming Europe*, ed. M. Cowles et al. Cornell University Press.

Cowles, M.G., Caporaso, J., and Risse, T., eds. 2001. *Transforming Europe: Europeanization and Domestic Change*. Cornell University Press.

Crisen, S.A.-M., and Carmin, J., eds. 2002. *EU Enlargement and Environmental Quality: Central and Eastern Europe and Beyond.* Woodrow Wilson International Center.

Cross, E. 1996. *Electric Utility Regulation in the European Union. A Country by Country Guide.* Wiley.

CzechInvest 1997. "Study of the Chemical Industry.".

Daly, H. 1995. "The Perils of Free Trade." In *Green Planet Blues*, ed. K. Conca et al. Westview.

Danchev, A. 1995. Enterprise Environmental Performance and Management in Bulgaria. Present Problems and Lessons from Western Experience. Working paper, Institute of Economics, Bulgarian Academy of Sciences, Sofia.

Darst, R. 2001. *Smokestack Diplomacy: Cooperation and Conflict in East-West Environmental Politics.* MIT Press.

Daskalov, R. 1998. "A Democracy Born in Pain." In *Bulgaria in Transition*, ed. J. Bell. Westview.

Dawisha, K., and Parrot, B., eds. 1997. *Authoritarianism and Democratization in Post-Communist Societies*. Cambridge University Press.

De Simone, L.D., and Popoff, F. 1997. *Eco-Efficiency: The Business Link to Sustainable Development*. MIT Press.

DeSombre, E. 2000. *Domestic Sources of International Environmental Policy: Industry, Environmentalists, and U.S. Power*. MIT Press.

Dimova, L. 1995. *Bulgarians on Environmental Problems*. Sofia: Corvette.

Duda, M. 1999. "New Electricity Tariff Setting System in Poland." *Biuletyn Energetyczny*, June: 11.

Duffield, J. 1992. "International Regimes and Alliance Behavior. Explaining NATO Conventional Force Levels." *International Organization* 46: 565–597.

Duina, F. 1999. *Harmonizing Europe. Nation States within the Common Market*. State University of New York Press.

Dujvelaar, C. 1996. Beyond Borders: East-East Cooperation among Environmental NGOs in Central and Eastern Europe. Report for Regional Environmental Center, Szentendre.

Dawson, A., and Fawn, R., eds. 2002. *The Changing Geopolitics of Eastern Europe*. Frank Cass.

EBRD (European Bank for Reconstruction and Development). 1992. Environmental Standards and Legislation in Western and Eastern Europe. Towards Harmonization. Bulgaria. Task IV: Air. Final Report.

Eichengreen, B., and Frieden, J., eds. 1998. *Forging an Integrated Europe*. University of Michigan Press.

Eising, R. 2002. "Policy Learning in Embedded Negotiations: Explaining EU Electricity Liberalization." *International Organization* 56, no. 1: 85–120.

Ekiert, G. 1996. *The State against Society: Political Crises and Their Aftermath in East Central Europe*. Princeton University Press.

EkoFinanse. 1997. "Ile Kosztuje Unia?" *Pazdziernik*: 7–10.

Eliste, P., and Fredriksson, P. 1998. "Does Open Trade Result in a Race to the Bottom? Cross-Country Evidence." Paper presented at World Bank conference on Trade, Global Policy and the Environment, Washington.

Energoproekt. 1996. Proekt za Normi na Emissii ot TEC i OC za SOx, NOx, Letjashta Neletliva Pepel i CO. Final Report. Sofia.

Enlarging the Environment. 1998. "Coping with Chemicals." Interview with G. Corcelle, Director of Chemicals Union of Directorate General for Environment of European Commission.

Enlarging the Environment. 1997. "New EU Acidification Initiative" Interview with Christer Agren, coordination of Acidification Strategy of European Commission. Brussels. Environmental Defense Fund. 1999. A Race to the Bottom: A Compilation of Export Credit and Investment Insurance Agency Case Studies.

Evans, P., Jacobson, H., and Putnam, R. 1993. *Double-Edged Diplomacy: International Bargaining and Domestic Politics*. University of California Press.

Droll, P. 2002. "The Big Push to Raise Standards." *Environment for Europeans* 9: 10–11.

European Council. 1988. Council Directive 88/609/EEC on Limitation of Emissions of Certain Pollutants into the Air from Large Combustion Plants. Also published in *European Community Environmental Legislation* (Office for Official Publications of European Communities, 1992).

European Council. 1993. "European Council of Copenhagen 21–22 June 1993. Conclusions of the Presidency." *Bulletin Quotidien Europe*, Document 1844/45.

European Council. 1996. Council Directive 96/92/EC of the European Parliament and of the Council of 19 December 1996 Concerning Common Rule for the Internal Market in Electricity. *Official Journal*, L 27, 30/01/1997.

European Dialogue. 1998. "Industry Helps with Adopting Complex Community Law."*European Dialogue* 6, November-December: 16–18.

Fagin, A. 1994. "Environment and Transition in the Czech Republic." *Environmental Politics* 3, no. 3: 479–494.

Fagin, A., and Jehlicka, P. 1998. "Sustainable Development in the Czech Republic: A Doomed Process?" *Environmental Politics* 7, no. 1: 113–128.

Faini, R., and Portes, R., eds. 1995. *European Union Trade with Eastern Europe: Adjustment and Opportunities*. London: Center for Economic Policy Research.

Federal Assembly of the Czech and Slovak Federal Republic. 1991. Transcript of Parliamentary Hearing on the Act on Clean Air from July 9, 1991, meeting 16. Statement by the Federal Minister Joseph Vavrousek.

Federal Committee of the Environment of the Czech and Slovak Federal Republic. 1992. National Report for the United Nations Conference on Environment and Development, Rio de Janeiro.

Federal Ministry of the Economy of the Czech and Slovak Federal Republic. 1991. The Principles of the State Energy Policy.

Financial Times. 1998. "Poland to Seek Waiver from Some EU Rules." March 6.

Finnemore, M. 1996. *National Interests in International Society*. Cornell University Press.

Fisher, D. 1993. "The Emergence of the Environmental Movement in Eastern Europe and Its Role in the Revolutions of 1989." In *Environmental Action in Eastern Europe*, ed. B. Jancar-Webster. M. E. Sharpe.

Frieden, J. 1991. *Debt, Development, and Democracy*. Princeton University Press.

Frieden, J., and Rogowski, R. 1996. "The Impact of the International Economy on National Policies: An Analytical Overview." In *Internationalization and Domestic Politics*, ed. R. Keohane and H. Milner. Cambridge University Press.

Frieden, J. 1999. "Actors and Preferences in International Relations." In *Strategic Choice in International Relations*, ed. D. Lake and R. Powell. Princeton University Press.

Gallup Poll London. 1990. Attitudes and Opinion of the Public Pertaining to the Social and Political Situation. Survey undertaken by the Gallup Poll London and Research Institute for Commerce on February 1990.

Garcia-Johnson, R. 2000. *Exporting Environmentalism: U.S. Multinational Chemical Corporations in Brazil and Mexico*. MIT Press.

Garrett, G. 1998. *Partisan Politics in the Global Economy*. Cambridge University Press.

Garrett, G., and Lange, P. 1996. "Internationalization, Institutions, and Political Change." In *Internationalization and Domestic Politics*, ed. R. Keohane and H. Milner. Cambridge University Press.

Garvey, T. 2002. "EU Enlargement: Is It Sustainable?" In *EU Enlargement and Environmental Quality*, ed. S. Crisen and J. Carmin. Woodrow Wilson International Center.

George, S. 1996. "The European Union. Approaches from International Relations." In *The European Union and National Industrial Policy*, ed. H. Kassim and A. Menon. Routledge.

Georgieva, K. 1993. "Environmental Policy in a Transition Economy: The Bulgarian Example." In *Environment and Democratic Transition*, ed. A. Vari and P. Tamas. Kluwer.

Georgieva, K., and Moore, J. 1997. "Bulgaria." In *The Environmental Challenges for Central European Economies in Transition*, ed. J. Klarer and B. Moldan. Wiley.

Giddens, A. 2000. *Runaway World: How Globalization is Reshaping Our Lives*. Routledge.

Glachant, M., ed. 2001. *Implementing European Environmental Policy. The Impacts of Directives in the Member States*. Elgar.

Glinski, P. 1996. *Polscy Zieloni. Ruch Spoleczny w Okresie Przemian*. Warszawa: IFIS PAN.

Glinski, P. 1998. "Polish Greens and Politics: A Social Movement in a Time of Transformation." In *Environmental Protection in Transition*, ed. J. Clark and D. Cole. Ashgate.

Goldstein, J. 1986. "The Political Economy of Trade: Institutions of Protection." *American Political Science Review* 80, no. 1: 161–184.

Gourevitch, P. 1986. *Politics in Hard Times. Comparative Responses to International Economic Crises*. Cornell University Press.

Gurowitz, A. 1999. "Mobilizing International Norms: Domestic Actors, Immigrants and the Japanese State." *World Politics* 51, no. 3: 413–445.

Grabbe, H., and Hughes, K. 1998. *Enlarging the EU Eastwards*. London: Royal Institute for International Affairs.

Grabbe, H. 2001. "How Does Europeanization Affect CEE Governance? Conditionality, Diffusion and Diversity." *Journal of European Public Policy* 8, no. 6: 1013–1031.

Grant, W. 1993. "Transnational Companies and Environmental Policy Making: The Trend of Globalization." In *European Integration and Environmental Policy*, ed. J. Liefferink. Belhaven.

Grant, W., Matthews, D., and Newell, P. 2000. *The Effectiveness of European Union Environmental Policy*. St. Martin's Press.

Green Circle. 1996. *Povez Mi, Kdo Je Nejkrasnejsi. Politicke Stany a Zivotni Prostredi*. Prague: The Green Circle.

Green Circle. 1997. Annual Report 1996.

GUS (Glowny Urzad Statystyczny). 1992. *Ochrona Srodowiska 1992*. Warsaw: ZWS.

GUS. 1993. *Ochrona Srodowiska 1993*. Warsaw: ZWS.

GUS. 1994. *Ochrona Srodowiska 1994*. Warsaw: ZWS.

GUS. 1995. *Ochrona Srodowiska 1995*. Warsaw: ZWS.

GUS. 1996. *Ochrona Srodowiska 1996*. Warsaw: ZWS.

GUS. 1997. *Ochrona Srodowiska 1997*. Warsaw: ZWS.

GUS. 1998. *Ochrona Srodowiska 1998*. Warsaw: ZWS.

GUS. 1999. *Ochrona Srodowiska 1999*. Warsaw: ZWS.

GUS. 2000. *Ochrona Srodowiska 2000*. Warsaw: ZWS.

GUS. 2001. *Ochrona Srodowiska 2001*. Warsaw: ZWS.

Haas, E. 1968. *The Uniting of Europe. Political, Social, and Economic Forces, 1950–1957*. Stanford University Press.

Haas, P. 1989. "Do Regimes matter? Epistemic Committees and Mediterranean Pollution Control." *International Organization* 43, no. 3: 377–403.

Haas, P. 1998. "Compliance with EU Directives. Insights from International Relations and Comparative Politics." *Journal of European Public Policy* 5, no. 1: 17–37.

Haas, P., Keohane, R., and Levy, M. 1993. *Institutions for the Earth. Sources of Effective Environmental Protection*. MIT Press.

Haggard, S., and Kaufman, R. 1995, eds. *The Political Economy of Democratic Transitions*. Princeton University Press.

Haggard, S., and Simmons, B. 1987. "Theories of International Regimes." *International Organization* 41: 491–517.

Hahn, R. 1990. "The Political Economy of Environmental Regulation: Towards a Unifying Framework." *Public Choice* 65: 21–47.

Haigh, N. 1990. *EEC Environmental Policy and Britain*. Harlow: Longman.

Hall, P. 1986. *Governing the Economy: The Politics of State Intervention in Britain and France*. Oxford University Press.

Halpern, L. 1995. "Comparative Advantage and Likely Trade Pattern of the CEECs." In *European Union Trade with Eastern Europe*, ed. R. Faini and R. Portes. London: Center for Economic Policy Research.

Hansen, W. 1990. "The International Trade Commission and the Politics of Protectionism." *American Political Science Review* 84, no. 1: 21–44.

Haverland, M. 2000. "National Adaptation to European Integration: The Importance of Institutional Veto Points." *Journal of Public Policy*, 20, no. 1: 83–103.

Held, D., et al. 1999. *Global Transformation: Politics, Economics, and Culture*. Stanford University Press.

Hellman, J. 1998. "Winners Take All: The Politics of Partial Reform in Post-communist Transition." *World Politics* 50, no. 2: 203–234.

Henderson, K. 1999. *Back to Europe: Central and Eastern Europe and the European Union*. UCL.

Heritier, A. 1999. *Policy Making and Diversity in Europe: Escape from Deadlock*. Cambridge University Press.

Heritier, A., et al. 2001. *Differential Europe. The European Union Impact on National Policymaking*. Rowman & Littlefield.

Hertzman, C. 1993. Environment and Health in Eastern Europe. Report for Environmental Action Program for Central and Eastern Europe. World Bank.

Hicks, B. 1996. *Environmental Politics in Poland. A Social Movement between Regime and Opposition*. Columbia University Press.

Hicks, B. 2002. "Setting Agendas and Shaping Activism: EU Influence on Central European Environmental Movements." In *EU Enlargement and Environmental Quality*, ed. S. Crisen and J. Carmin. Woodrow Wilson International Center.

Hoffmann, S. 1966. "Obstinate or Obsolete? The Fate of the Nation-State and the Case of Western Europe." *Daedelus* 95: 862–915.

Hoffmann, S., Keohane, R., and Nye, J., eds. 1993. *After the Cold War: International Institutions and State Strategies in Europe, 1989–1991*. Harvard University Press.

Holzinger, K., and Knoepfel, P., eds. 2000. *Environmental Policy in a European Union of Variable Geometry? The Challenge of the Next Enlargement*. Basel: Helbing & Lichtenhahn.

Hurrell, A., and Kingsbury, B., eds. 1992. *The International Politics of the Environment*. Clarendon.

ICCA (International Council of Chemical Associations). 1996. Responsible Care. Status Report.

Inotai, A. 1999. "Benefits and Costs of EU Enlargement: Theoretical and Practical Considerations on Trade Policy Issues." In *Governance, Equity and Global Market*, ed. J. Stiglitz and P. Muet. Oxford University Press.

Inotai, A. 2000. "The Czech Republic, Hungary, Poland, the Slovak Republic, and Slovenia." In *Winners and Losers of EU Integration*, ed. H. Tang. World Bank.

Institute for Sustainable Development. 1998. *Unia Europejska a Ochrona Srodowiska. Wybrane Fakty I Przemyslenia.* Warsaw: ISD.

Institute for Sustainable Development. 1999. *The Position Papers of the Polish Environmental Non-governmental Organizations on the Environmental Effects of Poland's Accession to the European Union.* Warsaw: ISD Publication.

Institute for Sustainable Development. 2000. Integracja Europejska. Pytania Organizacji Ekologicznych do Rzadu. Nie Tylko o Srodowisko UE?.

International Energy Agency. 1990. *Energy Policies: Poland. 1990 Survey.* OECD.

International Energy Agency. 1992. *Energy Policies: Czech and Slovak Federal Republic.* OECD.

International Energy Agency. 1994a. *Energy Policies of the Czech Republic.* OECD.

International Energy Agency. 1994b.*Poland. Energy Policies. 1994 Survey.* OECD.

International Energy Agency. 2001. *Czech Republic 2001 Review.* OECD/IEA.

Jakubczyk, Z. 1997. "Eco-Labeling Schemes in Poland." In *Eco-Labeling and International Trade*, ed. S. Zarrilli et al. St. Martin's Press.

Jancar-Webster, B. 1998. "Environmental Movement and Social Change in the Transition Countries." *Environmental Politics* 7, no. 1: 69–90.

Jacoby, W. 1999. "Priest and Penitent: The European Union as a Force in the Domestic Politics of Eastern Europe." *East European Constitutional Review* 8: 1–2.

Jankowski et al. 1998. Poland. Compliance with the European Union Air Emission Standards. Cost of Alternative Strategies for Reducing Sulfur Emissions. Final Report for World Bank.

Jehlicka, R., and Kostelecky, T. 1992. "The Development of the Czechoslovak Green Party since the 1990 Elections." *Environmental Politics* 1, no. 1: 72–94.

Jehlicka, P., and Tickle, A. 2002. "Environmental Policy and European Union Enlargement; A State-Centered Approach." In *EU Enlargement and Environmental Quality*, ed. S. Crisen and J. Carmin. Woodrow Wilson International Center.

Jendroska, J. 1998. "Environmental Law in Poland, 1989–1996: An Assessment of Past Reforms and Future Prospects." In *Environmental Protection in Transition*, ed. J. Clark and D. Cole. Ashgate.

Jha, V., Markandya, A., and Vossenaar, R. 1999. *Reconciling Trade and the Environment: Lessons from Case Studies in Developing Countries*. Elgar.

Johnson, L., et al. 2000. "New Chemical Notification Laws in Japan, the United States and the European Union." In *Regulatory Encounters*, ed. R. Kagan and L. Axelrad. University of California Press.

Kabala, S. 1993. "The History of Environmental Protection in Poland and the Growth of Awareness and Activism." In *Environmental Action in Eastern*, ed. B. Jancar-Webster. Sharpe.

Kalinova, E., and Baeva, I. 2000. *Bulgarskite Prehodi 1944–1999*. Sofia: TILIA.

Karaczun, Z. 1993a. Policy of Air Protection in Poland. Part I (till December, 1991) and Part II (the changes in 1992). Warsaw: Institute for Sustainable Development.

Karaczun, Z. 1993b. Policy of Air Protection in Poland. Part III—January–December 1993. Warsaw: Institute for Sustainable Development.

Karaczun, Z. 1996. Policy of Air Protection in Poland. Part IV—January 1994–December 1996. Warsaw: Institute for Sustainable Development.

Karasimeonov, G., ed. 1997. *The 1990 Election to the Bulgarian Grand National Assembly and the 1991 Election to the Bulgarian National Assembly. Analyses, Documents and Data*. Berlin: Rainer Bohn Verlag.

Katzenstein, P. 1978. *Between Power and Plenty. Foreign Economic Policies of Advanced Industrial States*. University of Wisconsin Press.

Katzenstein, P. 1985. *Small States in World Markets: Industrial Policy in Europe*. Cornell University Press.

Katzenstein, P. 1996. *The Culture of National Security: Norms and Identity in World Politics*. Columbia University Press.

Keck, M., and Sikkink, K. 1998. *Activists Beyond Borders. Advocacy Networks in International Politics*. Cornell University Press.

Kelley, J. 2001. Norms and Membership Conditionality. The Role of European Institutions in Ethnic Politics in Latvia, Estonia, Slovakia and Romania. Ph.D. dissertation, Harvard University.

Keohane, N., Revesz, R., and Stavins, R. 1996. The Positive Political Economy of Instrument Choice in Environmental Policy. Working paper, Belfer Center for Science and International Affairs, Harvard University.

Keohane, R., and Levy, M., eds. 1996. *Institutions for Environmental Aid. Pitfalls and Promise*. MIT Press.

Keohane, R., and Milner, H., eds. 1996. *Internationalization and Domestic Politics*. Cambridge University Press.

Keohane, R., and Nye, J. 1972. *Transnational Relations and World Politics*. Harvard University Press.

Keohane, R., and Nye, J. 2000. "Globalization: What's New? What's Not? (And So What?)." *Foreign Policy* 118: 104–119.

Kikuchi, R. 1997. Reciklirane na Vazdushnite Zamarsiteli ot Termichnite Centrali v Marica Iztok: Avtoreferet. Ph.D. dissertation, HTI, Sofia.

King, G., Keohane, R., and Verba, S. 1994. *Designing Social Inquiry. Scientific Inference in Qualitative Research*. Princeton University Press.

Kitschelt, H., et al., eds. 1999a. *Continuity and Change in Contemporary Capitalism*. Cambridge University Press.

Kitschelt, H., et al. 1999b. *Post-Communist Party Systems. Competition, Representation, and Inter-Party Cooperation*. Cambridge University Press.

Klarer, J., and Moldan, B., eds. 1997. *The Environmental Challenge for Central European Economies in Transition*. Wiley.

Klotz, A. 1995. *Norms in International Relations. The Struggle Against Apartheid*. Cornell University Press.

Klub Polskie Forum ISO 14000. 1996. *Polskie Forum ISO 14000 Bulletin*. Warsaw.

Klub Polskie Forum ISO 14000. 1997. *Information Bulletin*. Warsaw.

Knill, C., and Lehmkuhl, D. 1999. "How Europe Matters. Different Mechanisms of Europeanization." European Integration Online Papers 3, no. 7.

Komitet po Associirane Bulgaria-Evropeiski Sajuz. 2000. Shesto Zasedanie. Razdel Okolna Sreda. Report of sixth meeting of Committee for EU Association of Republic of Bulgaria.

Koulov, B. 1998. "Political Change and Environmental Policy." In *Bulgaria in Transition*, ed. J. Bell. Westview.

Krakow Academy of Economics. 1996. Developing of Cost Methodologies and Evaluation of Cost-Effective Strategies for Achieving Harmonization with EC Environmental Standards. Final Report for Ministry of Environmental Protection, Natural Resources and Forestry, Republic of Poland.

Kramer, J. 2002. "Enlargement and the Environment: Future Challenges." In *EU Enlargement and Environmental Quality*, ed. S. Crisen and J. Carmin. Woodrow Wilson International Center.

Levy, M. 1993a. "East-West Environmental Politics after 1989: The Case of Air Pollution." In *After the Cold War*, ed. S. Horrmann et al. Harvard University Press.

Levy, M. 1993b. "European Acid Rain: The Power of Tote-Board Diplomacy." In *Institutions for the Earth*, ed. P. Haas et al. MIT Press.

Levy, M. 1995. "International Cooperation to Combat Acid Rain." In *Green Globe Yearbook*.

Liberatore, A. 1992. "Towards Sustainability?" *Current Politics and Economics of Europe* 2, no. 4: 275–287.

Lipschutz, R. 1996. *Global Civil Society and Global Environmental Governance*. State University of New York Press.

Litfin, K. 1998. *The Greening of Sovereignty in World Politics*. MIT Press.

Lofdahl, C. 2002. *Environmental Impacts of Globalization and Trade. A Systems Study*. MIT Press.

Lubiewa-Wielezynsky, W. 1998. Harmonization of Polish Environment Law with EC Directives. Presentation of Industrial Chemistry Research Institute.

Luterbacher, U., and Sprinz, D. 2001. *International Relations and Global Climate Change*. MIT Press.

Lynch, D. 2000. "Closing the Deception Gap: Accession to the European Union and Environmental Standards in East Central Europe." *Journal of Environment and Development* 9, no. 4: 426–437.

Maillet, L. 1994. "New Air Rules Worry Czech Firms." *East/West Letter* 3, no. 4: 1–4.

Majone, G. 1996. *Regulating Europe*. Routledge.

Markova, H. 1996. Czech Environmental NGOs and Their Long-Term Financial Sustainability. Report for NATO Democratic Institutions Program.

Martin, L. 1992. *Coercive Cooperation. Explaining Multilateral Economic Sanction*. Princeton University Press.

Martin, L., and Simmons, B. 1998. "Theories and Empirical Studies of International Institutions." *International Organization* 52, no. 4: 729–759.

Martin, L., and Sikkink, K. 1993. "U.S. Policy and Human Rights in Argentina and Guatemala, 1973–1980." In *Double-Edged Diplomacy*, ed. P. Evans et al. University of California Press.

Martin, P. 1990. "The New Government." *RFE/RL Research Report*, July 27, 9–15.

Matev, N., and Nivov, N. 1997. "Implementation of Pollution Charges and Fines in Bulgaria." In *Theories and Methods*, ed. R. Bluffstone and B. Larson. Elgar.

Mayhew, A. 1998. *Recreating Europe. The European Union's Policy towards Central and Eastern Europe*. Cambridge University Press.

McCubbins, M., Noll, R., and Weingast, B. 1989. "Structure and Process, Politics and Policy: Administrative Arrangements and the Political Control of Agencies." *Virginia Law Review* 75: 431–485.

McGowan, R. 1993. *The Struggle for Power in Europe*. London: Royal Institute for International Affairs.

McNicholas, J., and Speck, S. 2000. Taxation on Energy and Transport: Domestic Policies in the Context of Climate Change in Central and Eastern Europe. Paper prepared for the World Resources Institute and the Regional Environmental Center.

Meacher, M. 1998. "Challenges Ahead for Candidates." *Enlarging the Environment*, Issue 11, July 1998, pp. 1–6.

Millard, F. 1999. *Polish Politics and Society*. Routledge.

Milner, H. 1988. *Resisting Protectionism. Global Industries and the Politics of International Trade*. Princeton University Press.

Ministry of the Economy of the Republic of Poland. 1997. *Fuels and Energy in 1996*. Warsaw: Energy Information Center.

Ministry of the Economy of the Republic of Poland. 2002. Polish Power Industry 2000.

Ministry of Energy and Energy Resources of the Republic of Bulgaria. 2002. Energy Strategy of the Republic of Bulgaria.

Ministry of Environment of the Czech Republic. 1991. *The Rainbow Program. Environmental Recovery Program for the Czech Republic*. Prague: Academia.

Ministry of Environment of the Czech Republic. 1995. State Environmental Policy. Document approved by the Government of the Czech Republic on August 23.

Ministry of Environment of the Czech Republic. 1996a. Report on the Environment in the Czech Republic in 1995.

Ministry of Environment of the Czech Republic. 1996b. *Statistical Environmental Yearbook of the Czech Republic 1996*. Prague: Czech Environmental Institute.

Ministry of Environment of the Czech Republic. 1997a. Draft Act on Chemical Substances and Preparations. Draft of June 30. English translation.

Ministry of Environment of the Czech Republic. 1999. *Statistical Environmental Yearbook of the Czech Republic 1999*.

Ministry of Environment of the Czech Republic. 2000. *Statistical Environmental Yearbook of the Czech Republic 2000*.

Ministry of Environment of the Czech Republic. 2001. *Statistical Environmental Yearbook of the Czech Republic 2001*.

Ministry of Environment, Natural Resources and Forestry of the Republic of Poland. 1990. Ordinance of the Ministry of Environmental Protection Natural Resources and Forestry of 12 February 1990 on Protection of Air against Pollution. English translation.

Ministry of Environment, Natural Resources and Forestry of the Republic of Poland. 1991. The National Environmental Policy of Poland.

Ministry of Environment, Natural Resources and Forestry of the Republic of Poland. 1995. Economic Instruments in the Environmental Management in Poland.

Ministry of Environment, Natural Resources and Forestry of the Republic of Poland. 1997a. Agenda 21 in Poland. Progress Report 1992–1996. Warsaw: National Foundation for Environmental Protection.

Ministry of Environment, Natural Resources and Forestry of the Republic of Poland. 1997b. Preparation of an Approximation Plan of Polish Legislation with the European Union Environmental Legislation. Final Report.

Ministry of Environment, Natural Resources and Forestry of the Republic of Poland. 1998a. "State of Approximation of the Polish Environmental Law to the

Environmental Law of the European Union." Presentation at PHARE environmental approximation facility (DISAE), Brussels.

Ministry of Environment, Natural Resources and Forestry of the Republic of Poland. 1998b. Dopuszczalne Wartosci Stezen Substancji Zanieczyszczajacych w Powietrzy. Dz. U. 98.55.355.

Ministry of Environment, Natural Resources and Forestry of the Republic of Poland. 1998c. Ordinance of the Minister of Environmental Protection, Natural Resources and Forestry on the Emission of Pollutants into the Air from Engineering Processes and Technical Operations. English translation. Original publication: Dz. U. 98.121.793, 22.

Ministry of Environment, Natural Resources and Forestry of the Republic of Poland. 1998d. Assumption to Ecological Strategy of Integration. National Program Preparing Poland to Membership in the Range of Environmental Protection. Document of the Department of International Relations.

Ministry of Environment, Natural Resources and Forestry of the Republic of Poland. 2000. The Triangle of Sustainable Development. Press release from the office of the Minister.

Ministry of Environment, Natural Resources and Forestry and Ministry of Industry and Trade of the Republic of Poland. 1996. Program Redukcji Emisji SO_2 w Energetyce Zawodowej.

Ministry of Environment, Natural Resources and Forestry, Ministry of Industry and Trade, and Ministry of Regional Planning and Construction of the Republic of Poland. 1996. Krajowy Program Redukcji Emisji SO_2 do Roku 2010.

Ministry of Environment and Waters of the Republic of Bulgaria. 1991. "Normi za Dopustimi Emissii (Koncentracii v Otpadachni Gazove) na Vredni Veshtesva, Izpuskani v Atmosferata." *Darzaven Vestnik* 81, 01/10/1991.

Ministry of Environment and Waters of the Republic of Bulgaria. 1997a. Letter of support for the CEFIC/PHARE Project on the Impact of the Commission's White Paper on the Chemical Industry in the CEE Countries.

Ministry of Environment and Waters of the Republic of Bulgaria. 1997b. *Bulletin* no. 2, November.

Ministry of Environment and Waters of the Republic of Bulgaria. 1998a. "Status of Environmental Approximation Activities." Presentation at fourth DISAE Seminar, Brussels.

Ministry of Environment and Waters of the Republic of Bulgaria. 1998b. Approximation of the Bulgarian Environmental Legislation with the Requirements of the EU Directives on Industrial Pollution. Project Description. Project financed by the Danish Agency of Environmental Protection.

Ministry of Environment and Waters of the Republic of Bulgaria. 1998c. Assistance in the Development of an Environmental Approximation and Training Program for Bulgaria. BUL-108. Final Report.

Ministry of Environment and Waters of the Republic of Bulgaria. 1998d. "Naredba No. 3 from 25/02/1998 za Usloviata i Reda za Utvarzdavane na

Vremenni Emissii na Vreni Veshtestva, Izpuskani v Atmosfernia Vazduh ot Nepodvizni Deistvashti Obekti, Svarzani s Nacionalnia Gorivno-Energien Balans na Stranata." *Darzaven Vestnik*, no. 51.

Ministry of Environment and Waters of the Republic of Bulgaria. 1999. Development of the Bulgarian Implementation Program for Approximation of EU Environmental Legislation. BUL 111. Project financed by DISAE.

Ministry of Environment and Waters of the Republic of Bulgaria. 2000. *Bulletin* no. 14.

Ministry of Environment and Waters of the Republic of Bulgaria. 2001. *Nacionalna Strategia po Okolna Sreda i Plan za Deistvie*. Sofia.

Ministry of Environment and Waters, Ministry of Health, Ministry of Industry, and Ministry of Regional Development of the Republic of Bulgaria. 1998. Nardba No. 2. Normi za Dopustimi Emissii (Koncentracii v Otpadachni Gazove) na Vredni Veshtestva, Izpuskani v Atmosfernia Vazduh ot Nepodvizni Iztochnici. *Darzaven Vestnik*, no. 51, 06/06/1998. Amended in 1999, *Darzaven Vestnik*, no. 34 (13/04/1999) and no. 73 (17/08/1999).

Ministry of Environment and Waters, Ministry of Health, Ministry of Industry, and Ministry of Regional Development of the Republic of Bulgaria. 1999. Nardba 15 ot 29/07/1999 za Normi za Dospustimi Emissii (Koncentracii na Otpadachni Gazove) na SO_2, NOx, i Prahoobrazni Veshtestva Izpuskani v Atmosfernia Vazduh ot Novi Golemi Gorivni Instalacii. *Darzaven Vestnik*, 17/08/1999.

Ministry of Health of the Republic of Bulgaria. 1997. Letter of Support for the CEFIC/PHARE project on the Impact of the Commission's White Paper on the Chemical Industry in the CEE countries.

Ministry of Health and Social Welfare of the Republic of Poland. 1996. Narodowy Program Zdrowia.

Ministry of Industry of the Republic of Bulgaria. 1997. Letter of support for CEFIC/PHARE project on the Impact of the Commission's White Paper on the Chemical Industry in the CEE countries.

Mitchell, R. 1994. *Intentional Oil Pollution at Sea: Environmental Policy and Treaty Compliance*. MIT Press.

Mol, A. 2001. *Globalization and Environmental Reform. The Ecological Modernization of the Global Economy*. MIT Press.

Moldan, B. 1997. "Czech Republic." In *The Environmental Challenge for Central European Economies in Transition*, ed. J. Klarer and B. Moldan. Wiley.

Moravcsik, A. 1991. "Negotiating the Single European Act: National Interests and Conventional Statecraft in the European Community." *International Organization* 45: 19–56.

Moravcsik, A. 1995. "Explaining International Human Rights Regimes. Liberal Theories and Western Europe." *European Journal of International Relations* 1: 157–189.

Moravcsik, A. 1998. *The Choice for Europe. Social Purpose and State Power from Messina to Maastricht.* Cornell University Press.

Murley, L., ed. 1995. *Clean Air around the World. National Approaches to Air Pollution Control.* Brighton: IUAPPA.

National Assembly of the Republic of Bulgaria. 1995. *Proektozakon za Chistotata Na Atmosfernia Vazduh.*

National Assembly of the Republic of Bulgaria. 1996a. Stenograma ot Parvo Chetene na Zakokonoproekta za Chistotata na Atmosfernia Vazduh from 26 January, 1996.

National Assembly of the Republic of Bulgaria. 1996b. Stenograma ot Vtoro Chetene na Zakokonoproekta za Chistotata na Atmosphernia Vazduh, from 15 May 1996.

National Assembly of Republic of Bulgaria. 2000. Stenograma of Vtoro Chetene na Zakokonoproekta za Zashtita ot Vrednoto Vazdeistvie na Himicheskite Veshtestva, Preparati i Producti.

National Center for Environment and Sustainable Development. 1990. *The State of the Environment in the Republic of Bulgaria. Annual Bulletin.* Sofia.

National Center for Environment and Sustainable Development. 1991. *The State of the Environment in the Republic of Bulgaria. Annual Bulletin.* Sofia.

National Center for Environment and Sustainable Development. 1992. *The State of the Environment in the Republic of Bulgaria. Annual Bulletin.* Sofia.

National Center for Environment and Sustainable Development. 1993. *The State of the Environment in the Republic of Bulgaria. Annual Bulletin.* Sofia.

National Center for Environment and Sustainable Development. 1994. *The State of the Environment in the Republic of Bulgaria. Annual Bulletin.* Sofia.

National Center for Environment and Sustainable Development. 1995. *The State of the Environment in the Republic of Bulgaria. Annual Bulletin.* Sofia.

National Center for Environment and Sustainable Development. 1996. *The State of the Environment in the Republic of Bulgaria. Annual Bulletin.* Sofia.

National Center for Environment and Sustainable Development. 1997. *The State of the Environment in the Republic of Bulgaria. Annual Bulletin.* Sofia.

National Center for Environment and Sustainable Development. 1998. *The State of the Environment in the Republic of Bulgaria. Annual Bulletin.* Sofia.

National Center for Environment and Sustainable Development. 1999. *The State of the Environment in the Republic of Bulgaria. Annual Bulletin.* Sofia.

National Center for Environment and Sustainable Development. 2000. *The State of the Environment in the Republic of Bulgaria. Annual Bulletin.* Sofia.

National Center for Environment and Sustainable Development. 2001. *The State of the Environment in the Republic of Bulgaria. Annual Bulletin.* Sofia.

National Fund for Environmental Protection and Water Management of the Republic of Poland. 1999. Ten Years of the National Fund for Environmental Protection and Water Management 1989–1999.

National Report of the Republic of Bulgaria on the Reduction of SO_2 and Dust Emissions. 1998. Proceedings from the Workshop of the Sofia Initiative on Local Air Quality, Bratislava.

National Statistical Institute of the Republic of Bulgaria. 1999. *Statistical Yearbook*. Sofia.

NEK (National Electricity Company). 1996. Annual Report.

NEK. 1998. Annual Report.

NEK. 1999. Annual Report.

NEK. 2000. Prestruktorirane na Nacionalnata Electricheska Kompania.

Natov, B. 1997. Ekologichnka Politica na Bulgarskata Stopanska Kamara—Politika na Ustoichivo Razvitie. Advisory paper of the Bulgarian Industry Association. Sofia.

Nicoll, W., and Schoenberg, R., eds. 1998. *Europe Beyond 2000. The Enlargement of the European Union towards the East*. Whurr.

Nowicki, M. 1997. "Poland." In *The Environmental Challenge for Central European Economies in Transition*, ed. J. Klarer and B. Moldan. Wiley.

Nye, J., and Donahue, J., eds. 2000. *Governance in a Globalizing World*. Brookings Institution Press.

OECD (Organization for Economic Cooperation and Development). 1995. *Environmental Performance Reviews. Poland*.

OECD. 1996. *Environmental Performance Review. Bulgaria*.

OECD. 1998. *Swapping Debt for the Environment. The Polish EcoFund*.

OECD. 1999a. *Cleaner Production Centers in Central and Eastern Europe and the New Independent States*.

OECD. 1999b. *Environment in the Transition to a Market Economy. Progress in Central and Eastern Europe and the New Independent States*.

Oye, K., and Maxwell, J. 1995. "Self-Interest and Environmental Management." In *Local Commons and Global Interdependence*, ed. R. Keohane and E. Ostrom. Sage.

O'Neill, K. 2000. *Waste Trading Among Rich Nations: Building a New Theory of Environmental Regulation*. MIT Press.

PCCI (Polish Chamber of Chemical Industry). 1994. The Polish Chamber of Chemical Industry (information bulletin).

PCCI. 1997a. *The Chemical Industry in Poland. 1996 Yearly Report*.

PCCI. 1997b. *Raport Srodowiskowy*. Warsaw.

PCCI. 1998. Impact of the Commission's White Paper on the Chemical Industry in CEE Countries. Country Report: Poland. Brussels: CEFIC-PHARE.

PCCI. 2002a. Chemical Industry in Statistics. Results of Industry in the First Half of 2001.

PCCI. 2002b. *Raport Srodowiskowy*. Warsaw.

Pehe, I. 1990. "Civic Forum and Public against Violence after the Elections." *RFE/RL Research Report*, June 22: 10–16.

Penchovska, A., et al. 1997. *Spravochnik na Nepravitelstvenite Organizacii za Opazvane na Okolnata Sreda*. Sofia: REC.

PHARE Program. 2000. Annual Report.

Pierson, P. 1996. "The Path to European Integration: A Historical Insitutionalist Analysis." *Comparative Political Studies* 29, no. 2: 123–163.

Pollack, M. 1997. "Delegation, Agency, and Agenda Setting in the European Community." *International Organization* 51, no. 1: 99–134.

Polish Environmental Law Association. 1997. Project on the Co-operation between the Ministry of the Environmental Protection, Natural Resources and Forestry and ecological NGOs. Final Report.

PPGC (Polish Power Grid Company). 1997. *Polish National Power System 1996*.

PPGC. 1999a. Annual Report 1998.

PPGC. 1999b. *Polish National Power System: 1999*.

PPGC. 2000. Data on Emissions of Major Air Pollutants and the Installation of Pollution Abatement Equipment in the Power Sector.

PPGC. 2002. Data on Emissions of Major Air Pollutants and the Installation of Pollution Abatement Equipment in the Power Sector.

Posner, R. 1974. "Theories of Economic Regulation." *Bell Journal of Economics* 5: 335–358.

Preston, C. 1997. *Enlargement and Integration in the European Union*. Routledge.

Przeworski, A. 1991. *Democracy and the Market*. Cambridge University Press.

REC (Regional Environmental Center). 1993. *Environmental Funds in Central and Eastern Europe. Case Studies of Bulgaria, Czech Republic, Hungary, Poland, and Slovak Republic.*

REC. 1995. *Status of Public Participation Practices in Environmental Decision-making in Central and Eastern Europe.*

REC. 1996a. *Approximation of European Union Environmental Legislation. Case Studies of Bulgaria, the Czech Republic, Estonia, Hungary, Latvia, Lithuania, Poland, Romania, Slovak Republic and Slovenia.*

REC. 1996b. *NGO Directory.*

REC. 1998a. Cleaner Production in Central and Eastern Europe. Paper of the REC's Business and Environmental Program.

REC. 1998b. *Reduction of SO2 and Particulate Emissions. Synthesis Report. Sofia Initiative on Local Air Quality.*

REC. 2000. *Greener with Accession? Comparative Report on Political Perceptions of the EU Accession Process and the Environment in Hungary, FYR Macedonia, and Romania.*

REC. 2001. *Environmental Funds in Accession Countries.*

Redmond, J., and Rosenthal, G. 1998. *The Expanding European Union: Past, Present, Future.* Lynne Rienner.

Responsible Care Program Poland. 2002. Overview of the Responsible Care Program in Poland.

Risse-Kappen, T., ed. 1995. *Bringing Transnational Relations Back In: Non-State Actors, Domestic Structures and International Institutions.* Cambridge University Press.

Risse-Kappen, T., Ropp, S., and Sikkink, K. 1999. *The Power of Human Rights. International Norms and Domestic Change.* Cambridge University Press.

Rodrik, D. 1997. *Has Globalization Gone Too Far?* Washington: Institute for International Economics.

Rodrik, D. 1999. The New Global Economy and Developing Countries. Making Openness Work. Washington: Overseas Development Council.

Rogowski, R. 1987. "Trade and the Variety of Democratic Institutions." *International Organization* 41, no. 2: 203–223.

Rogowski, R. 1989. *Commerce and Coalitions.* Princeton University Press.

Ruggie, J. 1998a. *Constructing the World Polity. Essays in International Institutionalization.* Routledge.

Ruggie, J. 1998b. "What Makes the World Hang Together? Neo-Utilitarianism and the Social Constructivist Challenge." *International Organization* 52, no. 4: 855–886.

Ruggie, J. 2002. "Taking Embedded Liberalism Global. The Corporate Connection." Paper presented at annual convention of American Political Science Association, Boston.

Sachs, J. 1994. *Poland's Jump to the Market Economy.* MIT Press.

Schimmelfennig, F. 2001. "The Community Trap: Liberal Norms, Rhetorical Action, and the Eastern Enlargement of the European Union." *International Organization* 55, no. 1: 47–80.

Sedelmeier, U., and Wallace, H. 2000. "Eastern Enlargment." In *Policy Making in the European Union*, ed. H. Wallace and W. Wallace, fourth edition. Oxford University Press.

Shugart, M., and Carey, J. 1992. *Presidents and Assemblies: Constitutional Design and Electoral Dynamics.* Cambridge University Press.

Sir William Halcrow and Partners, Ltd. 1997. Environmental Legislation Gap Analysis for the Czech Republic. Report prepared for the Czech Ministry of the Environment and the European Commission.

Sissenich, B. 2002. "Transnational Networks and Limits of Effectiveness: EU Enlargement and Social Policy in Poland and Hungary." Paper presented at Workshop on Enlargement and European Governance, Turin.

Slaughter, A. 1997. "The Real New World Order." *Foreign Affairs* 76, no. 5: 183–197.

Smith, A. 2000. *The Return to Europe. The Reintegration of Eastern Europe into the European Economy*. St. Martin's Press.

Sobolewski, M., and Taylor, E. 1996. The Organizational Structure of Environmental Protection in Poland. Discussion paper, Regulatory Policy Research Center, Oxford.

Stark, D., and Bruszt, L. 1997. *Postsocialist Pathways: Transforming Politics and Property in East Central Europe*. Cambridge University Press.

State Inspectorate of Environmental Protection of the Republic of Poland. 1993. *State of Environment in Poland*.

State Inspectorate of Environmental Protection of the Republic of Poland. 1998. *State of Environment in Poland*.

Statute on Chemical Substances and Preparations. 11 January 2001. Republic of Poland. *Dz.U.* 2001, No. 11, item 84.

Steinberg, P. 2001. *Environmental Leadership in Developing Countries: Transnational Relations and Biodiversity Policy in Costa Rica and Bolivia*. MIT Press.

Steinoff, J. 1999. "On the Way to the Union." *Biuletyn Energetyczny*, June: 7.

Stigler, G.J. 1971. "The Theory of Economic Regulation." *Bell Journal of Economics and Management Science* 6: no. 2: 3–21.

Stiglitz, J. 2002. *Globalization and Its Discontents*. Norton.

Stone, R. 2002. *Lending Credibility: The International Monetary Fund and the Post-Communist Transition*. Princeton University Press.

Suchanek, Z. 1997a. CEMC and ISO 1400 in the Czech Republic. CEMC policy paper.

Tallberg, J. 2002. "Paths to Compliance: Enforcement, Management, and the European Union." *International Organization* 56, no. 3: 609–643.

Tang, H., ed. 2000. *Winners and Losers of EU Integration. Policy Issues for Central and Eastern Europe*. World Bank.

Taras, 1998. "Voters, Parties, and Leaders." In *Transition to Democracy in Poland*, ed. R. Staar. St. Martin's.

Tokarczuk, A. 2000. "Strategy of Coexistence." *Biuletyn Energetyczny*, June: 8–9.

Tsebelis, G. 1995. "Decision Making in Political Systems: Veto Players in Presidentialism, Parliamentarism, Multicameralism and Multipartyism." *British Journal of Political Science* 25: 289–325.

Tsebelis, G. 2002. *Veto Players: How Political Institutions Work*. Princeton University Press.

Tsebelis, G., and Garrett, G. 2001. "The Insitutional Foundations of Intergovernmentalism and Supranationalism in the European Union." *International Organization* 55, no. 2: 357–390.

UNECE (United Nations Economic Commission for Europe) 1993. Environment and Economics: an Assessment of the Situation in Bulgaria. Report submitted by delegation of Bulgaria.

UNECE. 1994. Protocol to the 1979 Convention on Lon-Range transboundary Air Pollution on Further Reduction of Sulphur Emissions. Oslo, June 14, 1994. 33 I.L.M. 1540.

UNECE. 1999a. New Air Pollution Protocol to Save Lives and the Environment. Press Release ECE/ENV/99/11, Geneva.

UNECE. 1999b. Protocol to Abate Acidification, Eutrophication, and Ground-level Ozone.

UNECE. 2000. Environmental Performance Review of Bulgaria.

Unia Wolnosci. 1997a. "Porozumienie Programowe Wyborczej Koalicji Liderow Ekologicznych oraz Unii Wlnosci." *Za Zjelona Alternatywa*, no. 16.

Unia Wolnosci. 1997b. *Wyborcza Koalicja Liderow Ekologicznych*. Pre-election bulletin of the Freedom Union.

Vachudova, M. 2001. "The Czech Republic: The Unexpected Force of International Commitments." In *Democratic Consolidation in Eastern Europe, Volume* 2, ed. J. Zielonka and A. Pravda. Oxford University Press.

Van Brabant, J. 1999. *Remaking Europe: The European Union and Transition Economies*. Rowman & Littlefied.

Vari, A., and Tamas, P. 1993. *Environment and Democratic Transition: Policy and Politics in Central and Eastern Europe*. Kluwer.

Victor, D. 1998. "Learning by Doing in the Nonbonding International Regime to Manage Trade in Hazardous Chemicals and Pesticides." In *The Implementation and Effectiveness of International Environmental Commitments*, ed. D. Victor. MIT Press.

Victor, D., et al., eds. 1998. *The Implementation and Effectiveness of International Environmental Commitments: Theory and Practice*. MIT Press.

Vlacek, M. 1992. "Energy and Environmental Policy Integration: A Survey of the Activities of the Czech Power Company." In *Conference on Energy and Environment in European Economies in Transition. Priorities and Opportunities for Co-operation and Integration*. OECD.

Vogel, D. 1995. *Trading Up: Consumer and Environmental Regulation in a Global Economy*. Harvard University Press.

Vogel, D. 1997. *Barriers or Benefits? Regulation in Transatlantic Trade*. Brookings Institution Press.

Vukina, T., Beghin, J., and Solakoglu, E. 1999. "Transition to Markets and the Environment: Effects of the Change in the Composition of Manufacturing Output." *Environment and Development Economics* 4: 583–598.

Wallace, H., and Wallace, W., eds. 1996. *Policy Making in the European Union*, fourth edition. Oxford University Press.

Waller, M. 1998. "Geopolitics and the Environment in Eastern Europe." *Environmental Politics* 7, no. 1: 29–52.

Warsaw Voice. 1999. "Financing a Greener Poland." No. 39 (750), September 26.

Weir, M., and Skocpol, T. 1985. "State Structure and the Possibilities for 'Keynsian' Responses to the Great Depression in Sweden, Britain, and the United States." In *Bringing the State Back In*, ed. P. Evans et al. Cambridge University Press.

World Environment Center. 1999. Waste Minimization and Energy Conservation Program.

World Bank. 1991. Poland: Environmental Management. Project Report.

World Bank. 1992a. Czech and Slovak Federal Republic. Energy Sector Review. January 28, 1992.

World Bank. 1992b. Czech and Slovak Federal Republic. Power and Environmental Improvement Project. Staff appraisal report, February 21.

World Bank. 1992c. Czech and Slovak Federal Republic. Joint Environmental Study.

World Bank. 1992d. Bulgaria. 1992 Energy Strategy Study. Report 10143–Bul.

World Bank. 1992e. Poland Environmental Strategy. World Bank report 9808–Pol.

World Bank. 1992f. International Trade and the Environment.

World Bank. 1994. Bulgaria. Environmental Strategy Study Update and Follow Up.

World Bank. 1996. Least Cost Approaches to Reducing Emissions of Acid Pollutants. Final Report prepared by Coopers and Lybrand.

World Bank. 1997. Poland. Country Economic Memorandum. Reform and Growth on the Road to the EU. Document of the World Bank 16858–Pol.

World Bank. 1998. Poland. Complying with EU Environmental Legislation. Final Report. Discussion Paper.

World Bank. 1999. Czech Republic towards EU Accession.

World Bank. 2001. Bulgaria—Country Economic Memorandum The Dual Challenge of Transition and Accession.

World Bank. 2002. World Development Indicators Online.

World Economy Research Institute. 2000. *Poland. International Economic Report 1999/2000*. Warsaw: ELIPSA.

Yandle, B. 1989. *The Political Limits of Environmental Regulation. Tracking the Unicorn*. Quorum Books.

Yarnal, B. 1995. "Bulgaria at a Crossroads. Environmental Impact of Socioeconomic Change." *Environment* 37, no. 10: 7–33.

Young, O. 1989a. *International Cooperation: Building Regimes for Natural Resources and the Environment.* Cornell University Press.

Young, O. 1989b. "The Politics of Regime Formation." *International Organization* 43, no. 3: 349–376.

Young, O. 1994. *International Governance: Protecting the Environment in a Stateless Society.* Cornell University Press.

Young, O., ed. 1998. *Global Governance. Drawing Insights from the Environmental Experience.* MIT Press.

Young, O., ed. 1999. *The Effectiveness of International Environmental Regimes: Causal Connections and Behavioral Mechanisms.* MIT Press.

Young, O. 2002. *The Institutional Dimensions of Environmental Change. Fit, Interplay and Scale.* MIT Press.

Young, O., and Osherenko, G., eds. 1993. *Polar Politics: Creating International Environmental Regimes.* Cornell University Press.

Zakon 157. 1998. Sb o Chemickych Latkach a Chemickych Pripravcich, a o Zmene Nektrych Dalsich Zakonu. Zdroj 54/1998 Sb. From 13.07.1998.

Zakon 352. 1999. Kterym se Meni Zakon c. 157/1998 SB., o Chemickych Latkach a Chemickych Pripravcich, a o Zmene Nektrych Dalsich Zakonu. Zdroj 111/199 SB.

Zakon za Chistotata na Atmosfernia Vazduh. 1996. *Darzaven Vestnik*, no. 45, 28/05/1996, amended *Darzaven Vestnik* no. 85, 26/09/1997 and *Darzaven Vestnik* no. 27 31/03/2000.

Zakon za Energetikata i Energiinata Effectivnost. *Darzaven Vestnik*, no. 64, 16/07/1999, amended *Darzaven Vestnik* no. 1, 4/01/2000.

Zakon za Zashtita ot Vrednoto Vazdeistvie na Himicheskite Veshtestva, Preparati i Producti. 2000. *Darzaven Vestnik*, no. 10, 02/04.

Zarrilli, S., et al., eds. 1997. *Eco-Labeling and International Trade.* St. Martin's Press.

Zielonka, J., and Pravda, A., eds. 2001. *Democratic Consolidation in Eastern Europe. Volume 2: International and Transnational Factors.* Oxford University Press.

Zylicz, T. 1993. Economic Instruments in Poland's Environmental Policy: Status and Prospects. Working Paper, Beijer International Institute of Ecological Economics, Royal Swedish Academy of Sciences, Stockholm.

Zylicz, T. 1998. Social and Economic Consequence of Approximation with the European Union in the Environmental Field. Advisory note to Ministry of the Environment Natural Resources and Forestry of Poland.

Zylicz, T., and Holzinger, K. 2000. "Environmental Policy in Poland and the Consequences of Approximation to the European Union." In *Environmental Policy in a European Union of Variable Geometry? The Challenge of the Next Enlargement*, ed. K. Holzinger and P. Knoepfel. Basel: Helbing & Lichtenhahn.

Index

Activism, 24–26, 69, 81, 82, 91, 100, 110, 122, 150, 159, 182

Bulgaria, 43, 44, 55–61, 66, 76–80, 153–182

Communism, 26, 42, 49, 65, 82, 99, 101, 107, 124, 126, 138, 149, 154, 155, 180, 186, 190
Cost of reform, 7, 22–24, 27, 82, 91
 in Bulgaria, 57, 59, 60, 163–166, 177, 181, 185
 in Czech Republic, 42, 104–107, 111, 112, 115, 120, 121
 in Poland, 53, 54, 123, 126–128, 136, 139–143, 146, 150
Czech Republic, 39–48, 66–72, 95–122

Ekoglasnost, 154–156, 160
Exportation
 of chemicals, 41, 48, 56, 62, 66
 of electricity, 93, 136, 161
 and environmental policy, 12–19, 45–47, 60, 77, 79

Health, 35, 73, 89–91, 95, 97, 99, 100, 102, 123, 131, 153, 154, 156, 169, 172–174, 181
Hydropower, 106, 135, 161

Legislation
 and air pollution, 87, 96, 101–109, 115–121, 127–133, 144–146, 169–178
 and chemical safety, 17, 35–38, 42, 45–47, 53, 54, 59–61, 65–84, 184, 185
 and energy, 109, 134–137, 162
 and environment, 3, 7, 157–160

Nuclear energy, 26, 106, 109–113, 122, 124, 135, 150, 156, 161–163

Poland, 47–55, 66, 72–76, 123–151
Privatization, 40, 41, 48–50, 56, 134

Responsible Care Program, 39, 44, 45, 50, 51, 54, 58